SORROW'S LONG ROAD

SORROW'S LONG ROAD

THE SCIENCE OF GRIEF

BARBARA BLATCHLEY

Columbia University Press *New York*

Columbia University Press
Publishers Since 1893
New York Chichester, West Sussex
cup.columbia.edu

Cataloging-in-Publication Data is available from the Library of Congress.
ISBN 9780231214926 (hardback)
ISBN 9780231222501 (trade paperback)
ISBN 9780231560535 (ebook)
LCCN 2025009796

Printed and bound by CPI Group (UK) Ltd, Croydon, CR0 4YY

Cover design: Julia Kushnirsky
Cover photograph: Jeanne Jackson

GPSR Authorized Representative: Easy Access System Europe,
Mustamäe tee 50, 10621 Tallinn, Estonia, gpsr.requests@easproject.com

This is for Christopher

CONTENTS

PREFACE

On January 11, 2018, my husband and partner of thirty-six years died. In that moment, lasting no longer than the snap of a finger, my life changed utterly. Everything became "before" and "after." One second he was there, and in the next he was forever gone.

The confusion and disorientation his death created were overwhelming. I couldn't seem to stay on track with any of the tasks in front of me. I couldn't remember what I was supposed to be doing. I wonder now if this might be what dementia feels like. I would find myself standing in a room, feeling that I had come into that room looking for something but completely unable to pull to mind what I was looking for. Was I looking for him?

There were so many things to do, and so many of them were things I'd never done before. My journal stopped being a record of my feelings and thoughts and for a while after he died became the place where I tracked or tried to track what needed to be done, the first of what would be many lists.

In the beginning, those lists included many questions. He had said he wanted to be cremated, so I needed to make arrangements for that. They asked if I wanted to watch, an image that completely creeped me out—I just couldn't do it. Should I have agreed? Did I dishonor him by not being able to watch? They asked me to pick out something to put his ashes in. I never realized there was such a selection, from ornate to plain. Walking through the "show room," I heard a voice in my head, repeating, "No, no, no, no—I can't do this." But I did it, anyway, wondering what he would "like" in the way of an urn. Like? How is that even a question? He wouldn't "like" anything ever again.

I had to tell the people I worked with and my students what had happened, and I had to teach my classes. My first day back in front of the class I opened my mouth to speak and nothing came out. I gestured for the students to give me a moment, turned around and faced the blackboard, pushed back the tears that seemed omnipresent, took a deep breath, and then turned back around and started in. Descriptive statistics, yeah, they're important right now. That voice in my head was still chanting, "No, no, no, no, no." How am I going to be able to do this again and again?

I needed to arrange for a memorial service. Will people want food? I needed to pick out a place for his ashes and get a marker made for the cemetery. (I also bought the plot next to his for me when it's my turn to go.) Is this spot OK? I wonder if he'd like it (and here we go again with the "liking it" question). I needed to hire a lawyer to help with making sure the kids got their inheritance because he didn't leave a will (my first time hiring a lawyer for anything). Should I just sell the farm in Alabama and give the money to the kids, or should this be their decision?

Endless questions popped up, and I tried to answer them in my journal. How do you write an obituary? Do I tell his sisters-in-law? What do I tell his extended family? If everyone shows up for the memorial service, where will I put them? Where is the Social Security office, and do I need to make an appointment there? I can't find his paperwork, his insurance, his pension information. How do I handle that? Where is the title for his truck?

There is, I discovered, some comfort in making lists and in eventually checking those things off the list. I liked the feeling of having gotten something done. I'm not sure if the comfort just came from the lists, however. I think that the real comfort, for me, was in writing again, even if it was just a list.

Christopher had been ill with cancer and then heart disease and getting steadily worse for several years before his death. We dealt with the spiraling appointments, pills, "procedures" (it's not surgery; it's a procedure), and treatment options as best we could, but they were a time suck, occupying more and more of our consciousness and more and more of our "free" time. About a year and a half before he died, the textbook on statistics I had written finally went to the copy editors, and the long, drawn-out process of getting it ready for publication began. Because I like to be busy, I started writing a new book on why people believe in luck. Chris, as he always did, was acting as my in-house editor. I tried to be diligent about things such as daily word count and to keep to a schedule but found it almost impossible. When he died, I was in the middle of a paragraph in the middle of chapter 5 of the book. I just stopped. I didn't write another word for this book for another year.

I did, however, write in my journal. It began with the lists and then became a sort of free-form, stream-of-consciousness effort to put down on paper what was happening to me in my head and in my heart.

Those journal entries made me feel better. Calmer, more organized, less of the hot mess I could see myself becoming.

I had experienced grief before. My father died when I was twenty-six, and it was awful. I remember seeing him in his casket, the stitches that held his mouth shut visible to the mourners who walked by, and crying so hard I was momentarily afraid I wouldn't be able to stop. Losing Christopher was worse by far. Those tears I didn't even try to stop. My father was the central pillar of my childhood, and his loss left a gaping hole. Christopher was my adulthood, what felt like the central pillar of everything. Chris had been there when my father died, helping me deal with his death. I was sixty-two when Chris died, and now there was no one to help anymore.

The statistics textbook came out the day after Christopher died. I didn't even notice. I finished the luck book and sent it off to my agent, only moderately interested in whether it would be published, mostly just proud of myself for having finished it (it was published). And then I found myself in a familiar place. Finished with one project and thinking about the next one. I started on a book idea that Chris and I had discussed doing together on the way neuroscience is depicted in the movies. I wrote a proposal for it and received some interest, but I also knew that *I* had zero interest in following up on it. In the back of my brain, all the questions about grief and grieving were spinning around. What I really wanted to do was to write about them. I wanted to use the pain and confusion I had written about in my journal and the research on grief and pain and loss that others had done to find my way to answers to those questions. I sat down and wrote the first personal story ("Losing You") in one go, burst into tears, and didn't write anything else for a solid week. But at the end of the week, I realized that I could do this, that writing about his death helped, so I kept going.

What started as a largely selfish, or at least self-focused, attempt to help myself became this book.

I hope what I have discovered about the incredibly difficult and painful task of grieving the loss of someone you love will help readers in their own journey.

May 27, 2024

SORROW'S LONG ROAD

1

THE BEGINNING OF GRIEF

Give sorrow words: the grief that does not speak
Whispers the o'er-fraught heart and bids it break.
—William Shakespeare, *Macbeth*

DAY 1: LOSING YOU

You were upstairs, lying down, trying to sleep. That day was a "cancer day." That morning, we had been to the oncologist for your chemo shot. Isn't it strange that the medicine holding the beast at bay made you feel so god-awful—nauseated, tired, and weak? I went to work and was able to think about other things, normal life, everyday ordinary success and troubles. You went home and felt crappy.

When I got home, you asked me to lie down with you for a while. I probably looked exasperated for a second or so. I had brought work home with me (as usual) and felt the need to get to it. And truth be told, I was still a little angry with you from the night before. We had argued. You had not been able to find a comfortable position in bed, tossing and turning and out of sorts. I suggested you could try propping up the pillows and sleep sitting up, and your reaction surprised me. You were angry at the suggestion, and I didn't understand why. I thought it would help if you were propped up a bit. I remember thinking, "I have to go to work tomorrow, and I can't sleep with you rumbling around in the bed. I just want you to sleep so that I can sleep." I was still annoyed and tired the next day.

But I did lie down with you, and I'm eternally grateful that I did. I rubbed your back, hoping to help you go to sleep. I also nursed my resentment a bit. You had been downright mean the night before, saying I wasn't sympathetic to your discomfort and that I didn't care about you. A small and petty part of me silently said, "See? I'm as sympathetic as can be here." We waste our time on the stupidest things. It wasn't until much later that I realized you were probably afraid. Did you know? Did I miss it?

I went downstairs and stood at the kitchen sink, looking out the window and thinking, "Tomorrow we have an appointment with the new (next) cardiologist. He'll have some news, some idea of what to do. We just have to get to tomorrow." I started in on the pile of papers that needed to be graded. I was trying to figure out how to direct a student to the right answer without telling her what the answer was when I heard it. A thump, something knocking into something else. Like the book you were reading dropping to the floor as you fell asleep. I called up to see if everything was OK. No answer. I figured you must be asleep, and I didn't want to bother you if it really had been just an accidental bump, so I didn't think any more about it. About thirty minutes later, I put the papers aside and went upstairs to see if you wanted any dinner.

You were lying on your side. In the dim light of the bedroom, you looked comfortable and asleep. If you were sleeping, I didn't want to wake you. I knew that chemo days were awful, but I thought that you hadn't eaten all day and might like some soup.

I called your name. No answer. I stepped in closer and called your name again. Nothing. I grabbed your foot and shook you. Nothing. I thought you were joking around and shook you harder. Still nothing. I was scared now. I grabbed your hand—still warm—and shook it hard. I remember thinking that if you were joking around, I was going to be seriously angry. "This isn't funny, Dude." (I called him "Dude" because *The Big Lebowski* was his favorite movie.) I turned on the lights so I could see your face. You looked peaceful. You looked asleep.

Except for your eyes. They were halfway open, and when I looked at them, I stopped breathing. Your eyes were wrong. There was no light in them. They were not that beautiful green anymore. They were flat and black and wrong. I read somewhere that when we die, our eyes stop making tears. That's how your eyes looked. Lifeless, flat, black, empty, dull, dry, and dead wrong.

You were gone.

I grabbed the phone and dialed 911, but I knew it was too late. I knew it was too late when the 911 operator told me to get you onto the floor so the EMTs could do CPR. I knew when she told me not to worry about being gentle in getting you off the bed—she said, "You can't hurt him now." I knew it was too late when they laid you out on the living-room floor so they had more room. I knew it was too late when I sat halfway down the stairs watching them use the defibrillator, forcing your poor tired heart to contract with each electric shock and then watching those waves fade away.

Was the noise that I didn't investigate the sound of you dying?

My grief starts with that moment. With the realization that I was too late: I've lost my connection with other people, and I'm afraid I'll never get another chance to have one.

THE TOPOGRAPHY OF GRIEF

I started writing this book because talking about grief helped. I am by no means the first or the only person to have suffered loss, nor will I be the last, although I wouldn't wish this on my worst enemy. Every human on the planet knows, at some level, that we all die. Some know it intellectually, having not yet experienced the devastation of loss. But, unfortunately, most of us know it intimately and emotionally and repeatedly. The older we get, the longer grows the list of those we've lost. I have a good friend who lost a son (a pain I cannot imagine), and I know others who have lost spouses or partners at a very young age, parents, grandparents, aunts, uncles, nephews and nieces, and every other kind of significant relation you can imagine. Loss and the grief that comes from it are a part of being human.

When I was a young child, both my maternal and paternal grandmothers passed away within a few years of each other. We children were not allowed to attend the funerals, and so my parents' reactions were for me a giant mystery. I wasn't sure what had happened to Grandmas Grace and Emma, and I was completely confused about why my mother was sobbing in the arms of our next-door neighbor and why my already taciturn father had abruptly become even more silent.

My father died when I was twenty-six and he was only fifty-seven. Dad's death was unexpected, although I suspect death, even when you can see it coming, is always unexpected. A good friend from graduate school died by suicide, and the guilt I felt for not noticing he was struggling lasted a very long time. Early in my teaching career, a former student was accidentally exposed to a very rare disease that was not diagnosed in time to save her. There were others as well—friends, students, relatives, each passing producing waves of sadness, confusion, guilt, anxiety, and even anger.

However, as painful as these losses and others were, the death of my husband surpassed them all in the depth, duration, and distress of the pain I felt. Studies have shown that the intensity of the pain experienced with loss as well as the duration of the worst of grief vary across relationships, but the emotional and physical response to loss is fairly consistent. For example, in studies of severity of stress across different kinds of losses, loss of a child is ranked highest, given that it is unexpected and is a violation of the normal course of life's events. Death of a spouse is ranked almost as high because we build our lives and our expectations around our partners, and losing that support is intensely painful. Loss of a sibling as an adult is also high on the stress list, as is loss of a parent, but both kinds of relationships are quite different from the ones we form with a life partner, and so their loss tends to be felt less intensely.[1] (I also have to say that I feel awful in making sweeping statements about how painful the loss of a given relation is. I see no particular profit in comparing pain.)

I started reading and doing research about a year after my husband died because I had questions about what was happening to me. My intention in writing everything down was to do what I usually do when I'm confronted by something I don't understand—I gather all the information I can find and try to wrestle it into a form that I can wrap my head around. I am not alone in my "find out everything I can about it" response to stress, disruption, and anxiety. It is popular enough to have earned its own name, *problem-focused coping*, in the scientific literature.

I'm a neuroscientist by training, so I also look to what the research on the grieving brain can tell us about both grief and the process of adapting to it. Research comforts me, and I had mountains of questions as I began the process of adjusting and adapting to the loss of my husband. My questions began with how the extraordinary pain of this experience could be in any way evolutionarily beneficial. Why do we have painful, negative, awful emotions? I wanted to know how long it would last—how many weeks, months, years would have to go by before I felt anywhere close to better. How would I ever be able to look forward, to focus on the future instead of the past? Was what I was going through "normal"? Does everyone feel this way? Why am I angry? Am I angry with him for leaving me? That's not right. He's the one who died. Does this anger make me a bad person?

There are many more such questions from both my own experience and from the experiences of others. I hope that I can offer some of the understanding I've garnered to others going through the same darkness. Lisa Shulman writes, "The experience of grief can overwhelm us but understanding the science behind grief can dispel mystery and restore a sense of control[;] . . . a better sense of control helps us find a path forward."[2] I agree, and I hope to offer at least some of that restored feeling of control to the reader.

Loss and the grief it elicits have always been integral to human life. As Jim Morrison put it, "No one gets out of here alive."[3] The audience for this book is everyone who has lost someone important to them and has been trying to navigate the upheaval that results. The COVID pandemic has highlighted grief and loss in a startling new way. As of October 2023, we had lost to COVID 1,136,920 people in the United States and 6,972,152 people throughout the world. Added to those figures was the always-growing number of the newly bereaved, making grief front and center in the news.[4] Dr. Arielle Schwartz says in her blog post "Grit, Grief, and Grace," "Having an understanding of the neuroscience of grief can illuminate ways we can navigate that complexity."[5] Providing some help in understanding what scientists and researchers have to say about loss and grief is my goal in this book.

GRIEF VERSUS GRIEVING

Let's start with some basic terminology. Many researchers make a distinction between *grief* and *grieving*, despite the fact that they are obviously related. Grief is usually understood to mean the emotional response we have to the loss of someone we love, the wide array of emotions we feel, such as sadness, anger, jealousy, and sorrow. Grieving, also called *mourning* or *bereavement*, consists of the outward-facing behaviors we engage in as we express that grief and the process we engage in as we adjust and adapt to that loss.[6] According to Sidney Zisook and Katherine Shear, two researchers who study grief and its treatment, the term *bereavement* or *grieving* refers to the "behavioral manifestations of grief, which are influenced by social and cultural rituals, such as funerals, visitations or other customs," and the term *grief* describes everything else: the emotional, cognitive, and behavioral responses we have to the loss of that significant relationship.[7]

John Archer, writing about grief and grieving, describes a model of grief used by Colin Murray Parkes, a psychiatrist and author of many books on grief, who says that our brain tries to maintain a stable model of the person we love and have bonded to and continually checks that model with input from the outside world. When there is a mismatch, a discrepancy between the input and the model because that person is no longer in the outside world, distress is created. This is grief. That distress aids us in searching for the person we love and can no longer find. This is grieving.[8]

It is sort of like saying that grief is the way our brains adjust to the fact that although we desire that the world around us be stable and reliably constant, it is not. And when we have demonstrable evidence that this desired world does not exist, we must adjust.

Grieving is a painful process just as grief is a cluster of painful feelings. But grieving allows us to express those painful grief feelings, and the pain will ease with time. The rituals we engage in as we mourn—going through funeral services, placing flowers on the grave, gathering with family and friends to memorialize the lost person—are created to help ease the pain of mourning and to provide us, the bereaved, with occasions to express our emotions and to help us find a way to continue without our beloved.[9] These rituals are intimately tied to our culture and the expectations of the society we live in.

GRIEF AND CULTURE

Because *grief* is universal, even if the ways we express that grief—the ways we *grieve*—vary widely across culture, there are many definitions and descriptions of these terms. Caroline Lloyd defines *bereavement* as "the loss . . . of a valued relationship[,] . . . [t]he feelings, thoughts, and actions we experience after the death of a significant attachment, and the reorganization and adjustment of our world without that person in it."[10] I like the completeness of this definition as it addresses not just the emotions associated with grief but also the process we go through in dealing with all that emotion. I'd like to add a somewhat simpler definition of *grief*: grief is an overwhelming, whole-body, and whole-mind response to the loss of someone we love.

Some features of grief seem to be common across culture and individuals, but everyone grieves in their own way and in their own time. The universality of grief has been the topic of study for sociologists and anthropologists interested in the ways culture and society influence us. Kelly Diane Meade studies the role of culture in grieving. She writes that "various feelings associated with and manifestations of grief are the same or similar regardless of race, ethnicity, culture or religion."[11] All humans love, and all humans experience loss. Grief crosses every line we have ever constructed between us.

Psychologists who study grief have detailed some of the cognitive, emotional, and physical symptoms that the grief-stricken experience. In terms of cognition, grief is characterized by loss of concentration, forgetfulness, slow thinking, and difficulty in making decisions, even very simple ones. In terms of emotion, in Western cultures we experience sadness, despair, loss of hope, helplessness, loss of control, fear, anxiety, and anger—basically a breakdown of our emotional equilibrium. Even our physical bodies are affected by loss and grief. We cry; we wander aimlessly; our immune system breaks down, making us more prone to illness; we have trouble sleeping; we're restless; and we engage in searching behaviors, looking for that lost person despite knowing that he or she is no longer there.

Lloyd writes that the grief each of us feels is unique because each of us is unique and has had a unique relationship with the deceased.[12] There is no standard checklist of symptoms, no standard set of grieving behaviors that you can work your way through. No "Thank goodness *that's* finally over" at the end. Many writers have said that there is no getting over grief, nothing akin to getting over an illness—now you're done with it, and you feel all better. Grief over the loss does not end; we just gradually learn to live with it. A great deal of the difficulty and pain in grieving involves trying to adjust to a new world without that person in it.

Grief can manifest itself in somewhat different forms. Sometimes we grieve an anticipated loss before the actual loss of that person even occurs. Grief usually takes a familiar, so-called typical form, appearing after the loss of an important other, following a painful course, but easing over time, much to the relief of the sufferers and their family and friends. Acceptance, recovery, and continuation with an altered but still satisfying and, we hope, happy life are possible in the long term with *normal grief.* However, it needs to be said that being able to "move on" does not mean that the feelings of sadness and yearning simply disappear. Because grief alters every aspect of life, I think of grief as being incorporated into that new postloss life. What grieving does is give us the tools to adapt to grief as a new part of life, in which we move on but are changed by the event we have experienced. I used to have a sign in my office that read, "Experience is a hard teacher because she gives the test first and the lesson afterwards." Just as all experiences change us, so does grief.

Again, I want to emphasize that the ways we express grief are heavily influenced by our culture. As Paul C. Rosenblatt explains in his contribution to the collected volume *Death and Bereavement Across Cultures* (2015), "Each culture has its own approaches to dealing with death, which almost always involve a core of understanding, spiritual beliefs, rituals, expectations, and etiquette."[13] Death and some form of grief may be universal, but there is nothing universal in the way grief is expressed, defined, and participated in or even in the way words such as *alive* and *dead* are interpreted. We should not expect that grieving in another culture will look just like grieving in our own.

TYPES OF GRIEF

Normal or Typical Grief

The grief that the majority of us experience is often classified as *typical grief*. I had a sort of visceral reaction to this characterization of grief, the kind of grief I was experiencing, when I first came across it, and I wasn't sure why until later. I think my reaction had to do with the way the word *typical* is often understood—minimal, ordinary, and average. It was like a diminishment of what I was feeling. My grief did not feel average or ordinary in any way. But then I started to read about other forms of grief and quickly changed my view. This understanding of *typical* is not what is intended by the label *typical grief*. The kind of grief I have experienced is typical in that the overwhelming changes it evoked were those that the vast majority of people (as many as 85 percent of all grievers) who lose a loved one also experience. It's sort of a mathematical sense of typical, if you will. Now *that* I can understand (I don't teach statistics for nothing). This book describes typical grief, the kind most bereaved experience. But there are other forms of the grief reaction to loss, and I want to describe them briefly.[14] Be aware as well that it is possible to experience more than one type of grief reaction. I see myself as a survivor of anticipatory grief as well as normal grief. I have been dealing with both forms. You might be as well.

Anticipatory Grief

In some sense, I suppose Chris and I were lucky in the way things ended. We had been fighting the good fight against cancer and heart disease for a bit more than twelve years when he died. Knowledge of the inevitable end was there; we both saw it. I think both of us had gotten used to ignoring it and just putting one foot in front of the other to go on as best we could. The treatments changed, each one failing eventually, but then there was a new one we could try. So we got used to the roller coaster ride from hope to despair and back to hope again. He died when we were hoping for yet another new treatment that would alleviate his suffering. This time the plunge into despair didn't resolve back into hope again because he was gone.

Shulman describes witnessing the decline and death of someone we love as "walking to the very edge of the end of life and peeking, even taking a small step, over the side. . . . As survivors, we aren't recovering from loss—we are lost. We wake up to a world we don't recognize, where entrenched habits and behaviors are obsolete."[15]

The grief that happens before the actual loss of the person we love is called *anticipatory grief*.[16] It resembles the grief we feel after the death of the loved one but with a couple of differences. First, there is often more anger in anticipatory grief than with grief after death. Second, this anger might be related to feelings of fear, which also characterize anticipatory grief: fear of the approaching death of someone we love, fear of having to abandon the plans we've made for the future, fear of losing financial stability, fear of losing the stability in our personal relationship with the dying person and in the family.

Researchers are divided on the question of whether anticipatory grief makes later after-death grief easier or harder to deal with. I think it likely depends on the person experiencing the grief. I remember feeling angry at the situation, furious at cancer as an adversary, and deeply afraid of the one and only outcome in front of us. I was also scared that feeling this way meant that I was giving up. I didn't want Chris to think that I thought things were hopeless, and so I didn't talk about it and tried hard not to think about it. The research has shown, however, that expressing these feelings of grief and anger and fear helps people cope with the stress of being the caretaker and with anticipatory-grief reactions, and I'm sure that's true. It is not, however, what I did.

I don't know if expected loss is any easier to deal with. I haven't suffered the completely unexpected loss of someone so integral to my life, so I have nothing to compare it with. I do know that, as contradictory as it may sound, even though I saw death coming, I still didn't expect it when it arrived.

Prolonged or Complicated Grief

Sometimes the grief someone experiences seems to deviate from the "normal" patterns. In a form of complicated grief called *inhibited grief*, the person may not seem to grieve at all, or in *delayed grief* there may be an unusually long delay in the expression of the longing, seeking, and crying behaviors usually seen immediately after the loss in normal grief. With *chronic grief*, the symptoms of grief may last much longer than they seem to in typical grief, or with *distorted grief* they are unusually intense or atypical. Researchers, clinicians, and grief counselors are focused on these distinctions among different types of grief reactions because there is a need to be able to distinguish between these less typical forms of grief and clinical depression. Depression and grief are not the same reaction (although grief can lead to depression), and they are usually treated differently. Researchers and those involved with helping the bereaved cope with their grief point out that depression and prolonged grief respond differently to treatment, so it would be a mistake to count them as the same reaction. They also point out that typical grief and complicated grief differ from each other not in the life-altering changes that loss creates but in terms of their severity, the distress they cause, and their persistence in the life of the bereaved after loss.[17]

From the moment that beloved other person ceases to be, grief begins, along with all of the behaviors and physical and psychological changes that come with it. I remember being restless and unable to sit still. The night Chris died, I wandered through the dark house picking things up and putting them down somewhere else. I still can't find the cup from the bathroom. I have no idea where I put it. I walked incessantly, pacing a groove in the living-room rug, sometimes returning from an hour-long walk through the neighborhood just to turn around and go back out again, two or even three times in a day.

On the day of his memorial service, I insisted that I drive my siblings to the cemetery because I needed to "do something normal." Nothing felt normal then; "normal" took a while to come back home. Worst of all, at least for me, was that sleep, which has always been difficult for me, became almost impossible. I would routinely get one or two hours of sleep and then awaken in the very early morning hours to stare at the ceiling and wait for the sun to come up.

The difference between what I thought the future would look like and what it looks like now is astonishing. I saw us growing old together, side by side, reading books in the sun and talking about them, spoiling his grandchildren. Now I see an abyss. If I had been asked what I wanted to say to God or the universe or whoever was listening, it would have been something like, "This isn't fair! We had plans; we had a future." It seemed that when he died, my future died with him. Philosophers tell us that we should not live life looking over our shoulder at the past, but on my worst days it felt as though that's all there was.

WHY DO WE GRIEVE?

About three months after Chris died, a friend who was also a recent widow sent me the following very poignant and accurate description of loss by Kelley Lynn:

> The death of a spouse or partner is different than other losses, in the sense that it literally changes every single thing in your world going forward. When your spouse dies, the way you eat changes. The way you watch TV changes. Your friend circle changes (or disappears entirely.) [*sic*] Your family dynamic/life changes (or disappears entirely). Your financial status changes. Your job situation changes. It affects your self-worth. Your self-esteem. Your confidence. Your rhythms. The way you breathe. Your mentality. Your brain function. (Ever heard the term "widow brain?" If you don't know what that is, count yourself as very lucky.) Your physical body. Your hobbies and interests. Your sense of security. Your sense of humor. Your sense of womanhood or manhood. EVERY. SINGLE. THING. CHANGES. You are handed a new life that you never asked for and that you don't particularly want. It is the hardest, most gut-wrenching, horrific, life-altering of things to live with.[18]

I can see all of these changes in my life. In between second one, when he was alive, and second two, when he died, every single thing in my life changed. I also read this description and found myself wondering how this overwhelming pain can be an evolutionarily good idea. Why do we all feel it? Why make ourselves more likely to die through vulnerability? It would seem to be the antithesis of what spurs the survival of the individual and the species.

Two theories of the evolutionary function of grief are cited most often in the literature. The first is called *reunion theory*. It is based on the research originally done by John Bowlby and more recently by John Archer.[19] Bowlby was a developmental psychologist in the 1950s studying what he described as the separation response in children. When a child is separated from their mother (or the primary caregiver, who is usually Mom), they exhibit a series of behaviors that Bowlby described as aimed at reuniting with the mother. The child cries (which has in the past brought the mother running); they search and wander, looking for her; they become hypervigilant, reacting to anything that might be the mother in the environment around them. All these reactions are aimed at reuniting with the mother. Bowlby noted that the grief response in adults looks very much like the separation response in children. Grief at the loss of a beloved partner, which an adult understands as a separation that will not have a reunion, is an inevitable result of the separation response we evolved as a species to keep important relationships and important people nearby.[20] The separation response is a part of us because when we're helpless children, it works. It increases the chances that we'll be reunited with the person we are bonded to. Because it is wired into us evolutionarily, it lingers into adulthood even though it cannot function to reunite anymore.

The other major theory of the evolutionary purpose of grief comes from Sigmund Freud and, more recently, Randolph Neese. The purpose of grief in this theory is to help us disengage from the deceased and reorganize our plans, priorities, and relationships so that we can cope with the drastic changes that death has created in our lives. Neese says that grief is "sadness that has been specialized to cope with the loss of a close relationship."[21]

Both the Bowlby/Archer and Freud/Neese theories have been combined into a *cognitive-evolutionary theory* of the purpose of grief. According to this explanation, in the early stages of dealing with the loss of a loved one we engage in behaviors that look like the classic separation response: we yearn for the lost person, we wander, we search, we're restless—all of the behaviors that evolved to hook us back up to that important person when we were children. As we realize the permanence of that loss, our grief shifts to be characterized by sadness, thoughts about the lost person, disengagement, reorientation, and reorganization.[22]

LOVE AND BONDING

I think we all grieve the loss of love because we are a social animal. We survive best when we bond to other members of our own species. If we are alone, death is much more likely to find us before it finds someone happily embedded in a group. Evolutionary biologists would argue that we are motivated to maintain the good of the group even when it puts our own survival at risk because maintaining the group helps ensure our own individual survival.

When we lose that bond, when we lose the person we love, we suffer that loss, exhibiting grief. The psychological, cognitive, and physical symptoms of grief are an alarm reaction in the body.[23] Our mind sees that something has gone seriously wrong in the world around us. An important link to others has been broken, and we engage in a slew of behaviors designed to deny the loss, to find that missing person and get them back, even though we know that getting them back is impossible. We also begin the slow and painful process of adjusting to that devastating loss.

Just as building the bond took time, so does adjusting ourselves to the loss of that bond. The process of adjustment is painful. The things we do when we grieve can look maladaptive, but they ultimately are not. They *are* adaptation; they are what adaptation looks like. Change, as we all know, can be very difficult, and as Kelley Lynn points out in her description of grief, this kind of change is unwanted, unasked for, and dreadful. When the very kind police officer who was part of the crew that responded to my 911 call for help gently led me to another room, hoping I wouldn't see the EMTs working on Chris (I looked anyway, fixated on the EKG screen, trying to will his heart to start up again; it was brutal to watch, but I couldn't look away), my first words to him were, "I don't want to be a widow!" This was not supposed to happen, and I pushed it away as hard as I could. Pointless, I know, but gut-level instinct took over. I did not want this.

These behaviors and feelings often come in waves, peaking and receding across time, sometimes changing from minute to minute or hour to hour in the beginning but gradually waxing and waning from day to day, even month to month. This variation is a blessing in some ways—imagine feeling nothing but despair and pain all day, every day. The wave pattern of strong emotion may be how our body and our brain protect us from devastation. Giving us a break from the pain, periodic surcease is likely essential to our ability to cope with the loss, even though the return of the pain when it peaks again can be awful in contrast.

And as Kelley Lynn writes, every aspect of life is changed by loss. We're adapting not just to change in one part of life but to change in *all* of life's parts. This kind of change is difficult, painful, and time consuming. For me, the initial pain has eased, but I have also discovered that this reworking of my life is not done and is not ever going to be finished. I am not and will not be the same person I was before my husband died. This is what I think most people mean when they say that grief is permanent. We are absolutely and completely changed by loss.

THE MIND, THE BRAIN, AND GRIEF

In this book, I explain what psychology and neuroscience have discovered about our painful emotional response to loss. In chapter 2, I describe what psychologists know about love and attachment as well as why we willingly engage in the often-risky behavior of getting and keeping love. I also explain what love does for and to our minds and brains. Chapter 3 examines the stress of grief and the changes it can create in our bodies. In chapter 4, I look at the emotions of grief and why behaviors such as crying and wandering and yearning for what is irretrievably lost persist for so long (it can feel like forever). Chapter 5 deals with pain, both psychological and physical, and the purpose that pain serves. And in chapters 6 and 7, I discuss adaptation, what both researchers and the bereaved mean by words such as *recovery* and *resilience*, and how humans go about making meaning from the loss as they struggle to move forward. Finally, in chapter 8 I detail what I have learned about grief and grieving and tell you the answers to my initial questions. I also ask others about their own personal experiences with loss and grief and what the process has been like for them. I am deeply indebted to everyone who agreed to talk with me about their losses. They are among the strongest people I know.

There are two seemingly disparate perspectives in this book. The first is based on my own personal experiences in coping with the loss of my husband and partner. These accounts come from my journal entries in the days and weeks following Christopher's death and detail the raw and emotional turmoil I was feeling at the time. A word of warning—these accounts at the beginning of the first few chapters might for some people trigger deep emotional responses. Be aware, please.

The second perspective comes from the very different world of science and research. I started reading the science in reaction to that personal emotional upheaval in part because I like science but also in part because researchers have noted time and again that finding out more about something that is stressful can help restore a sense of control over that event, in much the same way that turning to faith and religious beliefs can help restore confidence and control.[24] I read about what others had experienced, so similar to what I have gone through, and how the scientific research put that experience in perspective. It helped.

By way of example, let me offer this. Grief researchers have been looking for an explanation of why some forms of grief, such as prolonged or complicated grief, happen to some but not all the bereaved. The answer may lie in the ways that we use what are called networks in the brain.

In 2009, the U.S. National Institutes of Health launched the Human Connectome Project, aimed at creating a wiring diagram for the human brain. They wanted to understand how different regions in the brain, such as those devoted to processing visual information and others focused on processing sounds or emotions or memory, work together to handle the very complex situations we humans create for ourselves. In an effort to figure out how activity in the brain generates everything that we think, feel, and do, researchers across the world are trying to understand these wiring patterns, how both experience and genetics together form and alter the various networks of systems in our brains, how we use these networks singly and jointly.

These images of networks in the brain are the result of studies done using functional magnetic resonance imagery (fMRI) scans. An fMRI scan will provide the researcher with a map of the areas of the brain that are most active when we're performing a particular task or even when we're just sitting and thinking about nothing much in particular. We can literally watch the brain while it is working by measuring how the cells that make up the brain use oxygen. The harder the brain cells are working, the more blood flows to that brain region. Blood delivers oxygen to the cells in the body and brain, so more blood flow means more oxygen, which in turn means the cells are working harder in that region.

A recent study using fMRI brain scans of people grieving the loss of someone dear to them has suggested that people suffering from prolonged grief may be using a network within the brain that helps code for and create our emotional responses—differently, that is, from people who experience so-called normal grief.[25] Among the study's participants, parts of this emotion network—for example, the limbic network—were hyperactive in people suffering from prolonged grief compared to the activity seen in people experiencing typical grief. In addition, hyperactivity between the emotion network and the rest of the networks in the brain, such as the attention network (guiding how we pay attention to the world around us), tended to increase over time as grieving symptoms worsened. These researchers concluded that shared activity between these networks might provide clinicians with a way to predict who is likely to experience prolonged grief, which in turn might give therapists the opportunity to intervene early for those at-risk individuals.

Even though I don't suffer from prolonged grief, there was comfort in seeing a potential explanation for some of what I'd been experiencing. When I saw Christopher's face everywhere and in everyone, it helped to think of this response as my emotional and attentional networks working overtime, searching for what I'd lost, looking to restore him. I don't see him everywhere anymore, and I would wager that my own neural networks are less hypersensitive now, perhaps back to their preloss patterns of function. I hope so.

If you're reading this because you have lost your own someone, I'm very sorry for your loss. Wherever you are in the process of grieving, adapting to, and coming to terms with that loss, I hope this book helps.

2

LOVE, ATTACHMENT, AND GRIEF

DAY 30(?): WHO AM I NOW?

In the weeks after Chris died, I found myself doing many things differently. I couldn't remember what I was supposed to be doing, which class I was supposed to be teaching, what day of the week it was, so I began making notes for myself. Detailed yet disorganized lists of things I needed to remember. "Make coffee. Today is Monday so your first class is Statistics. Eat something before you go to work, don't forget your gloves, go to the gas station, the car is nearly empty."

We shared a thirty-six-year history of leaving notes for each other. A note from Christopher, written a few weeks before he died, is still on my desk. In the early days after he died, looking at it would make me cry, but I couldn't bear the thought of it being tossed away, so I framed it and looked at it, and I cried. Now it makes me smile. It's so very much "him" in every good way. He was just checking in; he used a scrap of paper that was lying around (there's another note on the other side), and he signed it with his usual self-portrait sketch emphasizing his curly hair and round face. It reads: "Went for a walk—Kroger at @ 4 pm-ish. Back soon. C-Mac." "Back soon" . . . there is no measure of how much I wish that were true.

FIGURE 2.1. A note from Chris letting me know where he'd gone, with his standard signature: "C-Mac" with a sketch of him (in case I forgot?).

In the beginning, I couldn't face this loss directly. I found other things to think about, to distract me from the overwhelming pain that thoughts of his death triggered. I went back to work fairly quickly because at work I didn't have to face it. At work, I had too many other things to think about and do. I needed to hold this pain at arm's length. It was always there when I got home. Sometimes I hesitated to open the front door because I knew that once I was inside, it would come crashing down on me. More than once, I found myself on my knees on the living-room floor.

The house was so empty, when before it had been filled with his presence. All of his things—his guitar, his books, photographs of him and us, CDs with the music he loved, his ratty old Georgetown sweatshirt from his college days that he wore just for banging around the house—everything was as it had been, except that he was not there. I began to turn the television on as soon as I got home, with the sound turned down low. It didn't really matter what was on because I wasn't really watching it. I just needed the sound of another human voice. I wanted the illusion of other people in the house.

Lisa Shulman, writing about how her life was radically changed by the death of her husband, echoes what I was feeling: "And with that, the altered life began: waking up each day in an unfamiliar world where all rules are scrambled. Where you can no more connect with yourself than with those around you. Where the default is a sense of alienation, not fitting in the world. Where a simple chat involves navigating land mines, and life's routine stories are fraught with emotion. Each utterance preceded by split-second analysis—Can I say this sentence without breaking down? Will the listener be too distressed to hear this?"[1]

I distinctly remember how when a friend at work turned to smile at me as I was approaching him, getting ready to ask me how I was, I burst into tears at the sight of him. He hadn't even said anything, I just cracked at the possibility that I would have to speak. I think I scared him. He told me he hadn't wanted to make me cry, and I tried to tell him that *he* didn't; it wasn't his fault, it wasn't anyone's fault—it was that Chris was gone.

Shulman writes about how condolence cards made her grief open again, about how they would tear her apart. I felt and still feel very much the same about the mail that continues to arrive for Chris. Every envelope addressed to him from Doctors Without Borders or UNICEF or another charity he supported is another knife. Didn't I tell them that he was gone? I thought I did. Maybe I didn't.

I remember a phone call from a very eager and persistent young man, anxious to complete a sale for some computer software that Chris had expressed interest in shortly before his death. I am fairly certain this same salesman had overheard me in a previous phone call advising Chris not to bother with the software he was selling. I didn't trust this young man, and it seemed a bad idea in general. In the phone call after Chris died, this man seemed unwilling to believe that Chris was gone, asking me very politely but very insistently (he called repeatedly, several days in a row) to "please put Christopher on the phone." Finally, after several days of saying, "He can't come to the phone right now," in utter exasperation I snapped, "I'd love to, but you're going to have to talk really loudly—he's dead." Did I really say that to this baffled young man? I think I did. When did I get this mean? (I don't know if he kept calling—shortly thereafter I turned off the phone). Shulman says that she missed the woman she was with her late husband, "I mourn the loss of that identity—that woman died with him."[2] The person I was when Chris was still here is no longer the person I am—that is undeniable.

Many writers have described grief as the price we pay for love. We find our other and invest our time and energy, our feelings and future, in that person. There's a price to this investment of self in another person, and that price is grief when the person dies. Without love there is no grief—we wouldn't need to grieve the loss of love if there were no love there to begin with.

Love is risky for many reasons (it might not be reciprocated, it might be invested in a person who won't value it, it might require more than we're able to give), but one of the greatest risks of bonding with someone else is that the loss of that bond is inevitable, and the breaking of that bond might be because of the horrible permanence of death. This risk seems to be the last thing most of us consider as we prepare to jump off that ledge into love. I know that it never occurred to me that Chris might die before me until well into our years together, even though he was thirteen years older. And yet, despite the risks, love seems to be a fundamental part of human nature and a connection with others that we actively seek out despite the dangers.

LOVE, ATTACHMENT, AND BONDING

All humans come equipped to experience emotional responses to the world, and the range of emotions we can experience turns out to be remarkably consistent across generations and cultures. However, defining just what exactly an emotion is can be tricky because everyone experiences emotions in their own unique and personal way and because differences in definitions abound owing to the ambiguity of language. Ask someone to define "love" or "grief," and you'll get as many different answers as people you asked. It can be like trying to define something like pornography, about which Judge Potter Stewart famously said that although he couldn't define it in so many words, "I'll know it when I see it."[3]

What Is Love?

The American Psychological Association online *Dictionary of Psychology* defines love as "a complex emotion involving strong feeling of affection and tenderness for the love object, pleasurable sensations in their presence, devotion to their well-being, and sensitivity to their reactions to oneself."[4] It sounds a bit cold, a bit clinical, involving no hearts and no flowers, but the definition is intentionally stark. The scientists studying this powerful human emotion and trying to pin it down need to be very clear about what it is and what it is not.

The APA definition goes on to say love comes in different forms, all featuring some version of a motivation to be close to the object of that love. Many researchers make a distinction between *passionate love*, where sexual desire and excitement are dominant, and *companionate love*, featuring relatively less passion but deep attachment, commitment, and intimacy.[5] Agreement among researchers on a precise definition of an emotion, such as love, does not yet exist, but not for want of trying to achieve it.

The approach that I like the best comes from the psychologist Robert Plutchik, who proposes that emotions are cognitions such as "knowing, learning and thinking" that have evolved in all animals, not just humans, to help us survive. Plutchik's "psychoevolutionary theory of emotion" proposes that emotions are evoked by specific events in the life of the person experiencing them when "issues of survival are raised in fact, or by implication."[6] The behaviors and physical responses that are part of our emotional response to the world are designed to reduce the threat to our survival and to restore balance, harmony, and safety.

Figure 2.2 graphically displays Plutchik's model. Some emotions are basic or core emotional responses, common to all humans. Plutchik defines eight of these fundamental emotions—anger, anticipation, joy, trust, fear, surprise, sadness, and disgust. They are shown on the individual spokes of his model, on the second ring from the center of the diagram. These core emotions are also arranged in opposing pairs so that graphically the

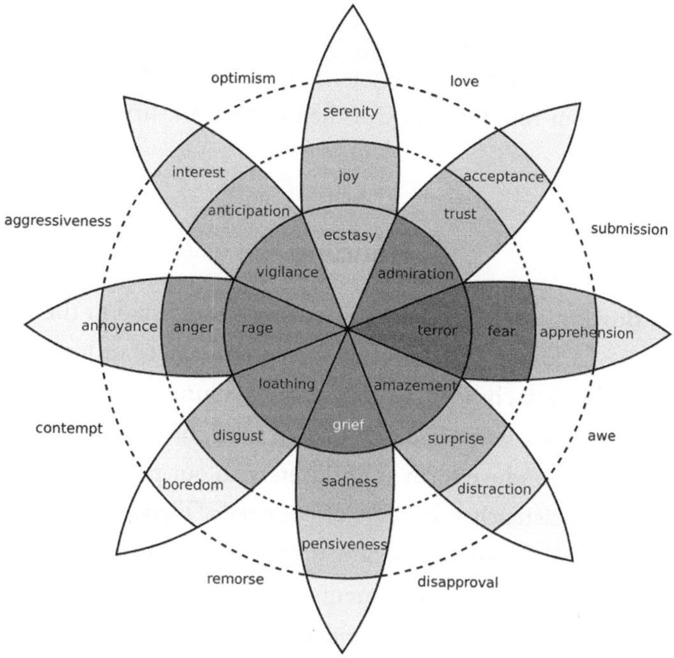

FIGURE 2.2. Plutchik's psychoevolutionary model of human emotions.
Source: Image from Wikimedia Commons, https://commons.wikimedia.org/wiki/File:Plutchik-wheel.svg, accessed March 13, 2025.

opposite of anger is fear, the opposite of sadness is joy, and so on. As you move toward the center of the emotion wheel, the emotions intensify. Notice that grief is on this "wheel" of emotions, just as love is. Grief must serve to help us survive. In this model, grief is the most intense form of sadness, and ecstasy is the most intense form of joy. If you move "outward" on an individual emotional spoke, you see weaker or milder forms of the emotion. For example, the milder form of sadness is pensiveness, and the milder form of joy is serenity. And some emotions, such as love are blends of core emotions. In Plutchik's model, love is a blend of trust and joy. All these emotional responses are essential to our survival as individuals and as a species. Both love and grief serve an evolutionary purpose. Both are wired into our brains and our minds because of their role in our survival.[7]

Bonding

John Bowlby, introduced in chapter 1, was interested in the formation of a loving bond and in the consequences of losing or lacking that attachment. In *Attachment*, volume 1 of his three-part series *Attachment and Loss* ([1969] 1982), he defines attachment as the bond we form as children with our parents or with whoever consistently takes care of our needs. That bond teaches us how to bond with others, how to form loving relationships as adults. We need those attachments in order to survive, so we are born equipped to attach or bond with other humans from the beginning of our lives.

Attachment figures provide us with stability, comfort, and safety as we learn to navigate the world on our own. Bowlby proposed that attachments were evolutionarily necessary, meaning they are essential to our survival because we are a social species. We need other members of our own species to survive. We form attachments to keep those others in our lives.[8]

Children use the person they're attached to as a safety net. Watch a toddler explore a new room with Mom seated behind her. She'll reach back to touch Mom and then move off to see what's new. That touch, that reassurance, allows her to explore. Without that touchstone, that essential marker of stability, the child grieves.

Bowlby describes two phases of the grieflike separation anxiety that children exhibit when faced with the absence of their caretaker. Children initially engage in the *protest stage* of separation anxiety. They are agitated, they cry, they wander aimlessly, and they are preoccupied with the lost person, pointing toward the door, asking for their mommy. All of these behaviors are also characteristic of the early stages of grief.

If the lost person does not return, the child's behavior changes to reflect what Bowlby refers to as the *despair stage* of separation. The restless wandering and searching behavior stop and are replaced by sadness, quiet, and lethargy. Again, these behaviors have reminded many researchers of adult grief at the loss of a loved one.

Bowlby saw both the protest and the despair phases as innate, having evolved to help us cope with the loss of a significant individual. The general arousal of the protest phase creates behaviors that force us up and out into the world and make reuniting with the lost person more likely. We search, we cry, we're loud and visible. The inhibition in the despair phase makes sense as a survival mechanism when the caretaker has still not returned. It makes us less visible and thus better able to avoid predators, and it conserves our energy for when we can't hide anymore.

Bowlby describes the attachments that we form in both childhood and adulthood as the strongest emotional responses we possess: "No form of behavior is accompanied by stronger feeling than is attachment behavior. The figures toward whom it is directed are loved and their advent is greeted with joy. . . . [A] threat of loss creates anxiety and actual loss sorrow; both moreover are likely to arouse anger."[9]

In volume 3 of his series, *Loss, Sadness, and Depression*, Bowlby also says that "loss of a loved person is one of the most intensely painful experiences any human can suffer[;] . . . to the bereaved, nothing but the return of the lost person can bring true comfort."[10] Even knowing that reunion with the lost person is never going to be possible does not stop the triggering of these behaviors. It makes no difference that adults know reunion is illogical and impossible. The response to loss is wired into our minds and brains, and so we express that loss as we begin to adapt.

Forming a bond with our caregiver when we are children and forming a bond with another person when we are adults help us learn to predict whether the protector will be there for us when we need them. That prediction feeds into our concept of self. Successfully bonded, we see ourselves as a person who can handle it when things go wrong. We learn to see ourselves as capable and strong. When you lose someone with whom you have built a relationship of trust and mutual support, you are abruptly back to being that child who can't find their mother. Now you need to learn how to do everything by yourself, and that is a shock for most of us.

The essence of attachment to another person has to do with trying to find comfort and support when we feel threatened and keeping that comfort nearby until the threat has passed. We learn to seek out the person we're attached to when we perceive threat or danger around us. We also learn that we can be successful in getting and keeping that care, and we incorporate that knowledge into what we know about ourselves. We feel rewarded, comforted, and loved. Our anxiety decreases, and we know that when we need to, we can elicit a supportive and caring response from the person we have bonded to.

When we are children, our primary caregiver (usually but not always our mother) provides us with what the psychologist Myron Hofer calls *biological regulators*.[11] Our caregiver, through the bond that we form with them, becomes a regulator of our physiology, our thinking, our expectations, and our emotions. The bond is a two-way street. They provide us with things such as body warmth, touch, and food. We learn to rely on them for comfort and nourishment, guidance and support, and we signal our demands to them, regulating their behavior, even before language is a possibility for us. This relationship and the bonds formed are reciprocal. The child shapes the caregiver's behavior as much as the caregiver shapes the child's behavior.

Hofer says that as adults we are still relying on the presence of special people in our environment to help us regulate our emotional and physical beings. Hofer's research has shown that the bonds we form with significant people in our lives—our parents and our romantic partners—are extremely important in keeping our physical bodies and our minds healthy and strong. When these bonds are broken, both mind and body are severely affected. These significant people have also become "regulators" of our adult behavior and our physiology. In the same way that sunlight regulates our biological clock, signaling us to be active during the daylight hours and inactive at night, the bond we form with our relationship partners comes to regulate many of the aspects of our daily lives.

BONDING, LOSS, AND GRIEF

When the person we love dies, the loss of that bond is often abrupt. The person who was the centerpiece of our lives is gone in an instant. All the links we've built between us, over days, weeks, and years are abruptly severed. All the habits we've developed that center around that significant person are suddenly without focus. That important other person was here; now that person is gone forever. I can attest to the fact that the change is disorienting as all hell. The author C. S. Lewis wrote movingly about his grief over the death of his wife, Helen (referred to in his book simply as "H"):

I think I am beginning to understand why grief feels like suspense. It comes from the frustration of so many impulses that had become habitual. Thought after thought, feeling after feeling, action after action had H. for their objective. Now their target is gone. I keep on through habit fitting an arrow to the string, then I remember and have to lay the bow down. So many roads lead thought to H. I set out on one of them. But now there's an impassable frontier post across it. So many roads once; now so many *culs de sac.*[12]

Grief is the name we give to the constellation of physical changes and emotions we feel when that connection is cut. My wandering through the house, picking things up and putting them down in different places, was me searching for my touchstone, my comfort, my "other." It was an innate behavior, a reaction to the threat to my survival that was activated by his abrupt disappearance from my life. I was not consciously searching for him; I knew intellectually that he was gone. That knowledge didn't stop the behavior, though; only time did that. It was a shock to realize how much I had come to depend on him, to shape what I did and thought around his presence.

Hofer also proposes two forms of grief he calls *acute* and *chronic.* [13] In acute grief, immediately after the loss, we experience waves of distress that peak and ebb and are evident in our physiology and our behavior. We feel physically weak, and we may sigh often and deeply (sighing has been described as a suppressed cry).[14]

In the chronic form of grief, the feeling of weakness persists, and changes in our cardiovascular, immunological, and endocrine systems occur, as well as weight loss or gain and sleep disturbances. The chronic form of grief is behaviorally characterized by lapses in attention and concentration, withdrawal from social situations, restlessness, anxiety, and the posture and expression of sadness. You may notice that the behaviors Hofer describes as part of the two forms of grief sound very much like Bowlby's two stages of separation anxiety. This similarity has not escaped notice in researchers interested in the emotion of grief. Bowlby's theory of attachment and the consequences of losing that attachment is often cited as a description of grief and grieving.

There is no timetable for grieving, no set amount of time it can be expected to last. The duration of grieving varies from person to person, from loving relationship to loving relationship. Remember the distinction I made earlier between grief and grieving. Grief is the emotional reaction to loss. Because that loss is permanent, so is the emotional response to that loss. Many writers say that the feeling of loss lasts forever.

Grieving is the process of adapting to that loss, the reconfiguring of our expectations and our lives. Grieving eases over time as we adapt to the reorganization of our lives. Mourning will fade even though the feeling of loss may never go away. This distinction makes sense to me. Think about the time it took to build the bond you shared with your beloved. Now think about how long it will take to rebuild life without that person. Grieving takes time because love takes time to create.

The researcher Phillipa Lally and her colleagues have suggested that it can take anywhere between 18 and 254 days to establish a habit.[15] To the extent that learning to rely on a particular person for our security and comfort when we feel threatened or endangered is a habit, it might take at least as long as that to learn that they are no longer here and that we no longer can rely on them. Other studies have shown that secure attachment develops between mother and infant during the first eighteen months of life.[16] It may well take at least as long to reorganize, revamp, and retune to the absence of a loved one. Mental resets are not easy to accomplish. They are painful and slow and sad, and they require that we adjust not only our emotions but also our physical brains.

MODELS OF GRIEF AND LOSS

When I was a child, I was fascinated by the model trains my father collected and even more intrigued by the train tracks, complete with villages, people, shops, bridges, and teeny-tiny trees, that he constructed in his basement workshop. He adored trains and airplanes, anything with a motor or an engine, especially if they were loud. We were probably the only house in town that had an enormous pen-and-ink drawing of a locomotive over the couch in our living room and replicas of fighter planes in attack formation dangling from our living-room ceiling.

The planes came in very handy one day when I asked my father about how big, heavy airplanes could stay in the sky. I got more than I bargained for as he used one of the models to explain air pressure and lift and the shape of the wings. Seeing the plane, being able to touch it, made the process much clearer. Most of us think about these sorts of things when we hear about models—physical replicas of much larger objects. Scientists also use models to help visualize and understand complex aspects of the world around us. Those models can be physical objects. For example, I have a model of the human brain in my office that comes in handy when I'm talking to students about neuroanatomy. But models can also be conceptual representations of very complex things such as human emotions (for example, Plutchik's diagrammatic model). There are also models of how we handle massive change in our lives, as we must do when we grieve.

Probably the most famous model of grief and grieving is the one proposed by the late psychiatrist Elisabeth Kübler-Ross. Kübler-Ross proposed that grieving happens in five stages: beginning with denial, then moving on to anger, depression, bargaining, and finally acceptance. The model describes a linear progression of movement through five phases of distress, ending with acceptance of the loss. This linearity suggested to many people that grieving individuals should deal with these aspects of loss one right after the other, handling each stage before moving on to the next.

There is, however, some indication that Kübler-Ross didn't intend to limit the model to just five stages and didn't intend the linearity of the model to be the most important take-home message, although that is what has happened. Kübler-Ross was attempting to get physicians and clinicians to start talking about the end of life with an eye toward helping people process both the loss of someone important in their lives and the eventual end of their own life (something that Western medicine still avoids talking about). Unfortunately, the message that many people have come away with in examining her model is that these stages are to be expected and that passage through them in a specific order is healthy and the best way to process grief.

Kübler-Ross wrote that these stages are not necessarily linear, not everyone experiences all of them, and they do not necessarily occur in a particular order. Rather, she wanted to describe the ways that many people experience grief and grieving, and her model, as good models should, got that conversation started.

Therapists and clinicians at the time Kübler-Ross published her book *On Death and Dying* (1969) adopted the stage idea, blending it with Sigmund Freud's proposal that the bereaved needed to engage in *grief work*. Grief work focused on breaking the attachment ties with the lost person so that the grieving person could adjust to their new life.[17] This view dominated clinical work in the 1970s, 1980s, and 1990s despite the growing realization among a number of therapists and researchers that stage models did not describe what actual bereaved people were telling them was going on.

More recent models of the grieving process have dropped altogether the idea of stages and linear movement through the stages. One of the more recent models, proposed by the researchers Margaret Stroebe and Henk Schut in 1999, is called the *dual process model*.[18] It is illustrated in figure 2.3.

The dual process model starts by describing the two types of stresses facing the bereaved; "loss-oriented" stress concentrates on the experience of the loss itself, and "restoration-oriented" stress coincides with the new tasks that now fall to the bereaved—having to adjust to doing things in a new way and to the new

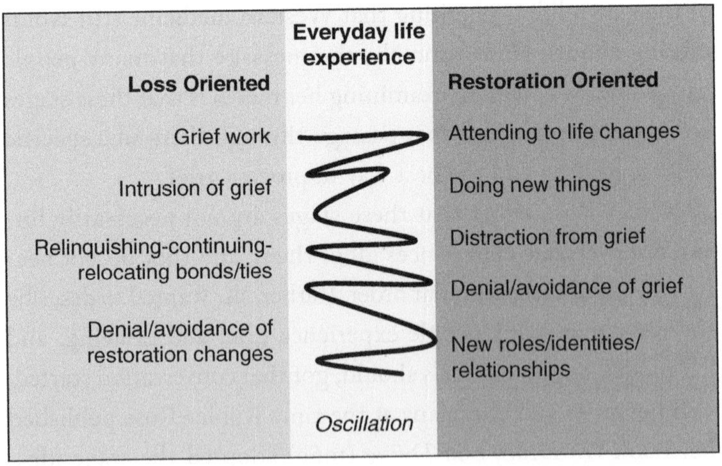

FIGURE 2.3. The dual process model of grief and grieving.

Source: Image from Wikimedia Commons, https://commons.wikimedia.org/wiki/File:The_dual_process_model_of_coping.png, accessed March 13, 2025.

role or roles now abruptly part of the grieving person's new life. Stroebe and Schut recognized that both stressors happen simultaneously and that both must be dealt with simultaneously. The wavy, oscillating line connecting these two stressors in the figure illustrates how people handle both of them, ping-ponging between dealing with the grief and dealing with a radically restructured life, sometimes attending to one while ignoring the other, sometimes trying to handle both at the same time. Coping with both types of stressors is embedded in everyday life, and both demand our attention and time.

One reason I like this model is that it beautifully describes the hot mess that grieving can be. We're not working our way through the process one step at a time as much as we are tackling what we can at any given point in the process, dealing with what our fractured emotions can handle, and avoiding what we cannot face at that moment.

LOVE AND GRIEF IN THE BRAIN

The brain is composed of cells, as is the rest of the body. Brain cells that send and receive messages are called *neurons*, and neurons work together in *circuits* or *networks* to create all the marvelous things we can do with our bodies and our minds. In chapter 1, I briefly introduced the idea of networks in the brain: neurons are linked together into organized circuits or networks, with each network processing different kinds of information. Experience with the world shapes the formation of those networks across our lifespan. Change in those experiences results in change in the neural circuitry that responds to those events.

There are many networks in the brain, processing information about the external world and sharing that information.

Sensory circuits convey incoming information from our senses, letting the brain know what is going on in the world; memory circuits are in charge of processing what is going on right now and what has happened to us in the past; emotion circuits create the extraordinary range of emotional responses we can have to the events unfolding in front of us; and motor circuits carry commands from the brain to the muscles attached to our skeletons, urging us to get up and move. There is even a circuit (perhaps more of a subcircuit belonging to the larger emotion circuit) that creates the feeling of love and bonding with another human being.

Understanding how the brain processes love can help us understand what happens in the brain when the object of our affection is taken away. Researchers have developed a number of neural models of both attachment formation and the role our attachment bonds have in the regulation of our bodies and minds.

Everything that makes up the two kinds of stressors in the dual process model is handled by the brain and the networks of nuclei and cortical regions that make it up. The *default-mode network* is a group of neurons that is active when we are *not* engaged in what's going on outside of ourselves but are instead focused internally. This network allows us to muse about the lost person during our downtime. The *salience* and *attention networks* direct our attention to what is important in the world around us, such as images and thoughts of that lost person. And the central executive network monitors everything all the brain networks are doing, and so it searches for distractions from our grief. These networks are responsible for both creating the feelings of love that develop as we bond with another person and for the behaviors so characteristic of grief and grieving when that person is lost.

Love Circuits in the Brain

Two interrelated circuits involved in the emotion of love have been identified using fMRI scans. One is a *cortical circuit*, made up of regions found within the cerebral cortex. The word *cortex* comes from the Latin word *cortice*, "bark (on a tree)." When we're talking about the brain, the cortex is the wrinkled outer surface of the brain just under the bony case of the skull. Researchers think the cortical circuit is involved in social cognition; the ways we think and feel about the social relationships we've formed as well as attention, memory, the links we've formed between memories, and representations of the self (our sense of who we are as individuals).[19]

Two main regions of the cortex are especially important parts of the cortical circuit: the *anterior cingulate cortex*, or ACC, which is toward the front of your brain, and the *insula*, which is deep in the human brain. Both the ACC and the insula play major roles in regulating our overall emotional response to the world, and both contribute to several of the networks I have just highlighted.

Several regions of the brain underneath the cortex are also involved with emotions, memory, and feeling motivated to go after what we want. Exactly how these subcortical regions are involved is still being debated, so I'll stick with the cortex for the moment. More about these regions is given in subsequent chapters.

The ACC, identified in the medial view of the brain on the bottom half of figure 2.4, is deep underneath the frontal lobe of the brain. This diagram shows what you might see if you were standing in the center of the brain, on the "midline," looking at the right half or hemisphere of the brain. The ACC is thought to be involved in making decisions about what to do next, especially decisions that concern our important social relationships. There is evidence that the ACC is important in our feelings of empathy when we see other people making mistakes or not being treated fairly, particularly when those people are folks we care about.[20] The ACC also has a role in directing our attention to important events in the world, such as seeing the face of someone we love, as well as memory and the way we understand our relationships with other people in our world (social cognition).

The insula (the Latin word for "island") is also found in the frontal lobe, deep underneath the outer layers of cortex, as shown in the lateral view of the brain on the top half of figure 2.4, where it is identified by the abbreviation AI, for *anterior insula*. It has a role in representing subjective feelings, including visceral feelings (such as that fluttery sensation we often feel when we see

---

FIGURE 2.4. The regions of the brain that process social relationships, emotions, motivation, and behavioral adaptations to change include the ACC, the PCC, and the insula. The lateral view in the top half of the figure shows the outer surface of the cortex of the left hemisphere of the brain. The regions of particular importance in this chapter are the anterior insula, depicted here as if it were on the surface of the brain but is actually underneath the outer layers of cortex, down deep inside folds of the cortex, and the ACC and PCC, shown in the bottom half. The bottom half shows a view from midline in the brain, looking at the right hemisphere. *Abbreviations*: ACC = anterior cingulate cortex, AI = anterior insula, AMY = amygdala, dIPFC = dorsolateral prefrontal cortex, EBA = extrastriate body area, FFA = fusiform face area, HTH = hypothalamus, mPFC = middle of prefrontal cortex, OFC = orbitofrontal cortex, PC = parietal cortex, PCC = posterior cingulate cortex, STS = superior temporal sulcus, v5 = visual region of the cortex, VS = ventral striatum, TPJ = temporal parietal junction, vPMC = ventral premotor cortex.

Source: Image from Wikimedia Commons, https://commons.wikimedia.org/wiki/File:Brain_areas_that_participate_in_social_processing.jpg, accessed March 13, 2025.

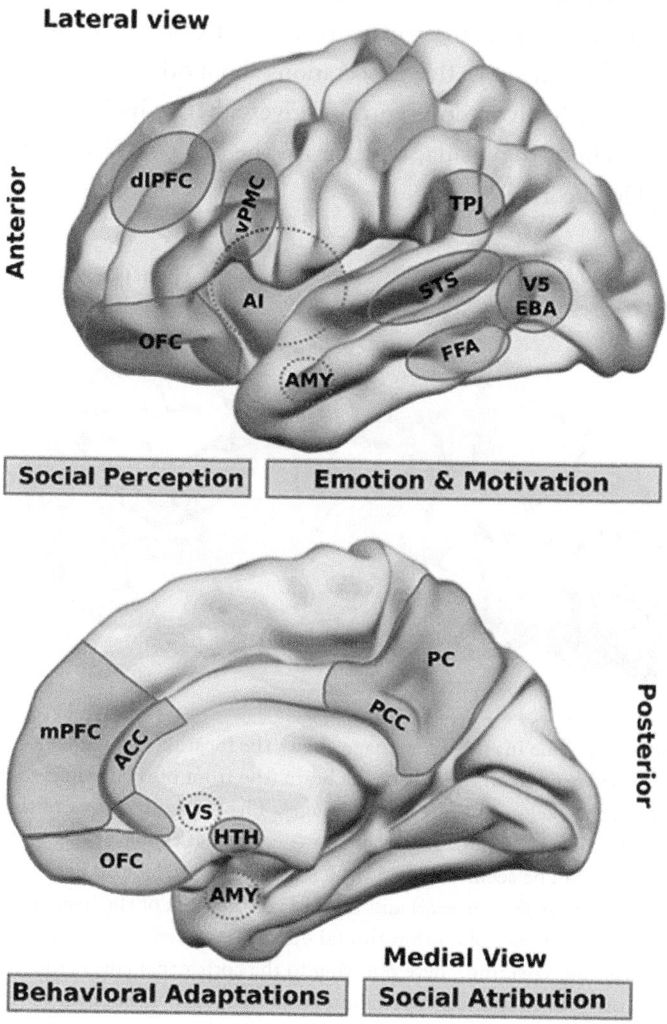
Lateral view
Anterior
dlPFC
vPMC
TPJ
AI
STS
V5
EBA
OFC
FFA
AMY
Social Perception
Emotion & Motivation
PC
Posterior
mPFC
PCC
ACC
VS
HTH
OFC
AMY
Medial View
Behavioral Adaptations
Social Atribution

our beloved), as well as in directing our attention, forming intentions, understanding the trustworthiness of other individuals, and forming our subjective expectations about other people and what they're likely to do. Figure 2.5 gives a better indication of the location of the insula, underneath the outer surface of the frontal lobe cortex.

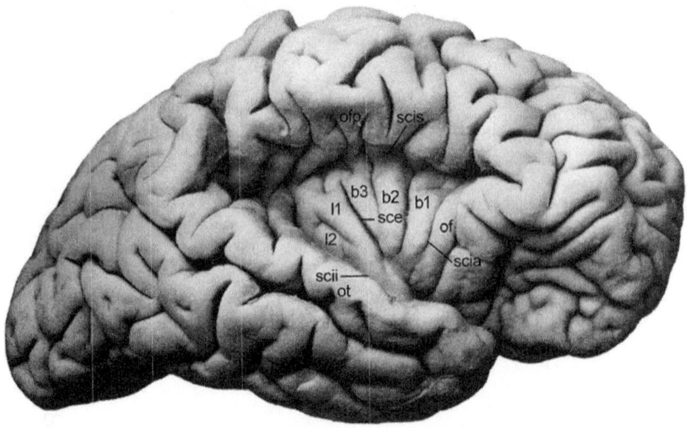

FIGURE 2.5. The insula. This image shows the location of the insular cortex, as viewed from the side of the brain (the front of the brain is to the right, the back to the left in this image). The overlying cortex of the frontal lobe has been pulled away or retracted so that the insula underneath can be seen. The folds of insular cortex are labeled b1 through b3 (*from front to back*) and I1 and I2 at the back of the insular cortex. *Abbreviations*: of = orbitofrontal operculum (*operculum* is a Latin term meaning "covering," here referring to the cortex that covers the insula), ofp = frontoparietal operculum, ot = temporal operculum, sce = central sulcus of the insula (a sulcus is a fold in cortical tissue), scia = anterior circular insular sulcus, scii = inferior circular insular sulcus, scis = superior circular insular sulcus.

Source: Image from Wikimedia Commons, https://commons.wikimedia.org/wiki/File:Human_Insular_Anatomy.png, accessed March 13, 2025.

Grief Circuits in the Brain

Researchers in 2003 asked grieving people to view an image of a lost beloved person while undergoing an fMRI scan.[21] These scans show activity in the cortical love circuitry, with some additional components. Both the ACC and the insula were activated (in the lab we say these regions "lit up" because on the computer-enhanced image color is added to active brain regions) when a grief-stricken study participant saw the image of the person they had lost. Both regions are involved in directing our attention toward a salient or important cue in the world. Grieving individuals are often preoccupied with images of the person they've lost, a sign of just how important the person is and was, so the image of the person lost becomes especially salient. In addition, the adjacent *posterior cingulate cortex* (PCC) (see figure 2.4), toward the back of your brain, also shows increased activity. The PCC's function is still being debated in neuroscience labs, but its central role in the default-mode network in the brain has suggested some possible functions for it.

The default-mode network is made up of a series of regions in the brain that work together. This network is activated when we're daydreaming, not really focused on any particular thing in the world around us. In fact, the default-mode network and the PCC are turned *off* when we're actively focused on accomplishing a specific task. This has led to the suggestion that the PCC might be involved in thinking that is directed internally, rather than outwardly, toward recollection of *autobiographical memories*, or memories of important events in our own individual life histories, or maybe just in daydreaming. Recent research has suggested that the PCC might also be involved in helping to control cognitive behaviors, perhaps even signaling the frontal lobes that something in the environment has changed and that, as a result, our behavior needs to change as well. Perhaps seeing the image of a deceased loved one activates memories of being with that person and signals an emergency at the loss of that person and the need to start the painful process of adaptation and change.[22]

Thus, the same circuits that we use to develop our feelings of love and attachment to another person are also involved in our feelings of grief and loss when that important person dies. The ACC directs our attention toward the person we love. The insula, modulating our expectations of other people and helping us form our intention to be with that person, and the subcortical regions involved in motivations, feelings of reward, and pleasure at being with that beloved person help to create our feelings of love and attachment. Those same regions are activated during grief and grieving, triggering the feelings of love but now without a live target for that powerful emotion, creating our preoccupation with images of our beloved and the things the loved person left behind (even something as innocuous as a note), and initiating memories of that person and the still strong motivation to be with them despite the impossibility of that desire. No wonder grief and grieving are so painful.

3

GRIEF AND STRESS

The Physiological Effects of Bereavement

THE FIRST THREE MONTHS

At some point in the days that followed Chris's death, I developed a persistent chest cold. I lost my voice, as I often do when I'm ill. The cold resulted in my croaking my way through classes when I returned to work. Ever tried to mime statistics? It's not easy. I turned down offers from friends to join me on my walks. I couldn't talk with them, anyway, and I wasn't sure I really wanted to. What on earth would I say?

The cold lasted several weeks. I would get better and then relapse, get my voice back and then lose it again. More bothersome was the feeling of being unable to sit still despite being tired. It is a very odd feeling to be both tired and restless, lethargic and jittery, at the same time. When I was a child, my father sometimes called me "Miss Fidget" because of my inability to sit still. I would sometimes watch TV upside down, sitting with my feet up the back of the chair and my head hanging down toward the floor, bouncing and jiggling my feet all the while. But that kind of restlessness passed as I grew older, and I learned how to be quiet. When Chris died, the fidgeting came back with a vengeance. Being still was once again almost impossible.

I walked incessantly, coughing and crying as I went, a wheezing, hacking mess. I couldn't be still long enough to watch a TV show all the way through; I jittered and bounced, getting up to walk and sitting back down to jiggle my foot, the urge to move relentless.

I didn't know if lethargy was the product of the illness or of the grief that came in waves, crashing down on me and making me feel pinned to the couch. Maybe it was both. The cold finally let up about three and a half weeks later. I was so relieved I noted in my journal that it was gone.

GRIEF AND THE STRESS RESPONSE

I didn't connect being ill with grieving until later in the process of dealing with Chris's passing. Now I realize that getting sick was likely a by-product of the stress that grieving produces. Researchers are vitally interested in the human stress response because stress is an omnipresent aspect of life. Stress is the reaction we have when something happens that threatens us or makes us feel pressured to act, usually when we're afraid that we can't handle that event or when we feel we have no control over what is happening.

The death of someone we love is one of the most stressful things that will happen to us in our lives, and it can trigger a profound disruption of our biological and psychological systems. Psychologists have designed several paper-and-pencil tests to assess our stress levels and our reactions to stressors in the world around us. These tests often consist of lists of common stressful events that the person taking the test checks off as being something that has recently happened.

For example, the Social Readjustment Rating Scale (also known as the Holmes and Rahe Stress Scale after the two psychiatrists who developed it to study the relationship between stress and illness) is one of the most commonly used measures. It lists forty-three common life events that you might have experienced in the past two years and rates them using a scale that ranges from 0 to 100, with higher numbers indicating greater stress levels. Death of a spouse rates a 100 on this scale, divorce a 73, the death of a close family member a 63, and the death of a close friend a 37. Not all stress-inducing events are "bad" events; even some happy events, such as adding a new family member (39 on the scale) or achieving something important (28) can cause stress (called "eustress," or good stress, because it is beneficial to the person experiencing it). Change for the better or for the worse, as it turns out, is stressful.

Grief tops these lists of potential stressors because it is caused by a drastic change in the life of the bereaved. The U.S. Institute of Medicine Committee for the Study of Health Consequences of the Stress of Bereavement points out that grief at the loss of someone we love is different from any other stressor. Unlike most other stressful events, such as the loss of a job or the breakup of a relationship, the bereaved suffer a loss that is forever. In addition, there is absolutely nothing we can do to reverse or undo that stressful event. The only way to bring back the lost love is via hallucinations or denial, neither of which is really effective. The committee goes on to say that the "central task for the bereaved . . . is to reconcile themselves to a situation that cannot be changed and over which they have no control."[1]

So, what exactly happens to us when we're stressed out? The stress response in humans is often referred to as the "fight-or-flight" response. When we're confronted by something that is stressful, we need to decide quickly how we're going to deal with it. There are two immediate options: we can run away from the stressor, or we can fight it. One response is not necessarily better than or more correct than the other. The circumstances of the stressful event as well as our past history of dealing with stress dictate which option we'll choose.

The brain and spinal cord, which make up the *central nervous system*, decide what to do next. Two components of the *peripheral nervous system*, made up of the nerves that carry information into and out of the central nervous system, are responsible for creating the fight-or-flight response—the urge to stand and fight or to run away. Those two components, the *sympathetic* and *parasympathetic divisions*, are shown in figure 3.1, detailing the components of the nervous system overall.

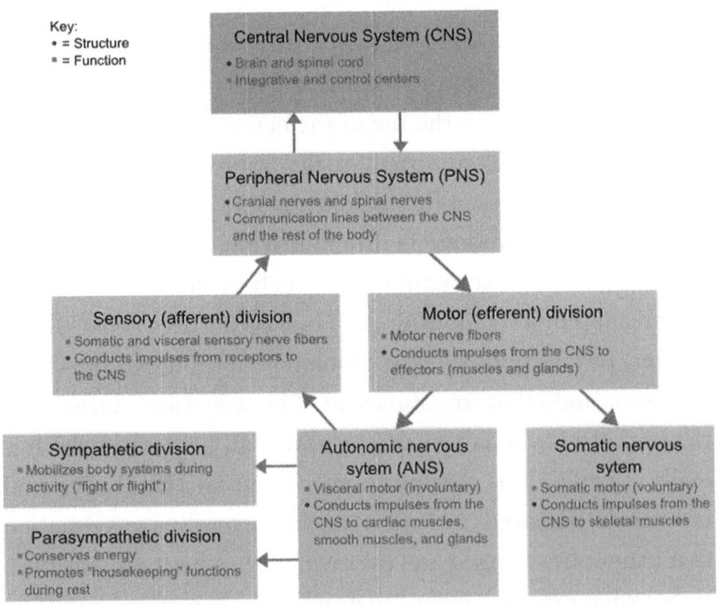

FIGURE 3.1. Diagram of the hierarchy of the nervous system in vertebrates. Within each box, the small dots indicate individual structures within the brain. The term *afferent* refers to information coming from the external world into the nervous system; *efferent* refers to information from the nervous system moving out to the body and the external world.

Source: Image from Wikimedia Commons, https://commons.wikimedia.org/wiki/File:NSdiagram.png, accessed March 13, 2025.

Both the sympathetic and the parasympathetic divisions of the peripheral nervous system communicate with the same muscles and glands, working together but having the opposite effects at their targets. The parasympathetic nervous system works to conserve energy, to maintain balance in the body. It is "in charge" when we're resting or not reacting to some dire event in the world. The sympathetic nervous system is activated when there's an emergency, when we need to do something right now (the "fight-or-flight" response). The sympathetic system gets our bodies ready to react to the threat.

Our physical and psychological systems are normally in balance, or a condition of *homeostasis*, and both body and mind work together to maintain that balance. Regulatory mechanisms are triggered when things such as body temperature, blood pressure, and the balance of salt and water in our bloodstream are not what they should be. The job of these regulatory mechanisms is first to detect the imbalance and then to reestablish homeostasis. Most of the mechanisms that act to reestablish balance in our physiology are automatic.

Threats to our survival throw that homeostasis out of whack and trigger defensive mechanisms that are designed to resist that threat or to get us away from it. We can deny that the threat is happening, we can come up with justifications for our own response or for the responses of another, or we can (sometimes) physically leave the scene of the stress. All are fight-or-flight responses. Grief is one such threat. Hofer says that "the central characteristic of [grief] is a failure of the normal smooth modulation and coordination of affect [emotion], behavior and physiological function into a stable daily pattern." In other words, the balance in our systems, our homeostasis, is disrupted by the loss of a person central to our lives—the loss of a biological regulator.[2] We react to this stressor, trying to restore balance in both body and mind.

Grief is more than a psychological-emotional response to loss. It also has effects on the way the body functions—that is, on our physiology—and those effects can have long-lasting consequences for our health. Loss is interpreted by the brain as an emergency, a threat to our continued survival. The person we have come to expect to be there, to rely on for support and help, is missing. In other emergencies, these fight-or-flight strategies usually work. In this particular emergency, though, they won't. But our knowing that they don't work won't prevent these autonomic and automatic responses from occurring.

THE HUMAN STRESS RESPONSE

Proposed by Hans Selye in 1946, the *general adaptation syndrome model* of the stress response says there are three stages to our response to an emergency.[3] When the emergency first happens, we go into the *alarm phase* (also known as the "fight-or-flight response") as we realize that there is a threat we need to react to. Our sensory systems send an emergency signal to the *amygdala*, part of the limbic, emotion-processing network in the brain. The amygdala then sends a signal to the *hypothalamus*, the command center for the *autonomic nervous system* in the brain, telling it that an emergency needs to be dealt with.

The hypothalamus then tells the *sympathetic division* of the autonomic nervous system to do what is necessary to make energy available so we can react to the threat. The *adrenal glands* (which sit on top of the kidneys) are told to release *adrenaline*, which makes the heart pump faster, which in turn makes more blood available to the muscles, so we can move quickly.

This chain of emergency-response centers in the brain and the body are collectively called the *hypothalamic-pituitary-adrenal*

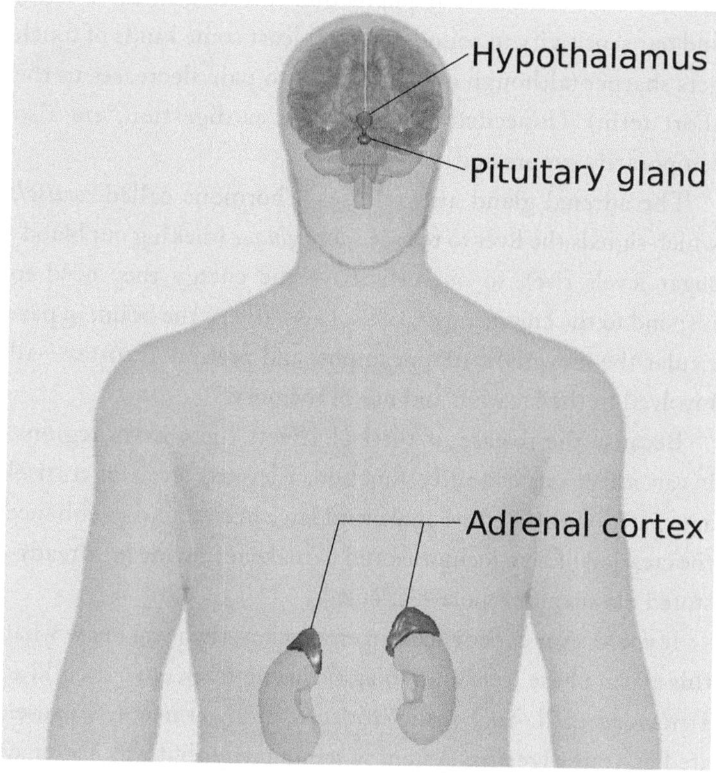

FIGURE 3.2. The hypothalamic-pituitary-adrenal gland axis.

Source: Image from Wikimedia Commons, https://commons.wikimedia.org/wiki/File:HPA-axis_-_anterior_view_(with_text).svg, accessed March 14, 2025.

axis. These disparate areas are linked together via the autonomic nervous system. Figure 3.2 shows the three main components of this axis that are activated in the alarm phase of the stress response.

Respiration rate also increases, making more oxygen available to muscles for quick movement and to the brain for an increase in alertness. Our pupils dilate, letting in more light, and our sensitivity to sound and to at least some kinds of touch gets sharper (although our sensitivity to pain decreases in the short term). Unneeded systems, such as digestion, are also temporarily suppressed.

The adrenal gland also releases a hormone called *cortisol*, which signals the liver to release more *glucose* (making our blood-sugar levels rise), so our cells have the energy they need to respond to the emergency. Cortisol also affects the brain, in particular the amygdala, hippocampus, and prefrontal cortex—all involved in the creation and use of memory.

Because the release of cortisol affects these brain regions, it can affect our cognitive function. Elevated levels of cortisol associated with an emotional event have been found to enhance the creation of new memories and to make retrieving an already-stored old memory more difficult.[4]

If you've ever experienced an emergency, then you know what this alarm phase feels like. Imagine losing track of your child at a crowded mall. Your stomach drops, and you may feel nauseated as your digestive system is temporarily shut off. You may start to sweat, your heart pounds, you start breathing faster, and your muscles tense. Your pupils dilate as you scan the crowd around you, frantically calculating how much time has passed since you last saw your child. When you spot your child watching the popcorn machine spew its contents into the bin, you feel the parasympathetic system kick in. Emergency handled, the body switches into restoration mode, and your heart rate comes back down, your breathing slows, and you may realize that the popcorn smells really good and you're hungry. You may also feel jittery for a bit because it takes a few minutes for the adrenaline levels to drop. It may take as long as thirty minutes for our bodies to return to homeostasis.

If you can resolve the emergency, you may never move on to the second and third stages of the general adaptation syndrome response. Unfortunately, when the emergency is the loss of a person we love to the permanence of death, there is no resolution possible, no quick restoration of homeostasis. If this happens, then the body enters what Selye called the *resistance phase* of the stress response.[5] In this stage, the sympathetic nervous system stays activated, and, pedal to the metal, we stay locked in the emergency.

There are consequences of long-term chronic stress for both the body and the mind. On the physical side, they include aches and pains, changes in our sleeping pattern (such as insomnia or oversleeping), changes in appetite, and lowered energy levels. On the psychological side, we may feel that we can't concentrate or that our thinking is fuzzy and cloudy (often referred to as "widow brain"). We shy away from social interactions and pull into ourselves emotionally.

Grief is a unique kind of stressor in that there is no way to "solve" the emergency. We may search for the lost person, but there is no way to find them. During the resistance phase, we are trying to adapt to the emergency, trying to change our thinking about the loss and the future and what we're going to do about it. Resolution of that stressor, coming to some kind of accommodation with it, may be as close to a solution to the emergency as we're going to get. That accommodation takes a while because we're reprogramming what we know about everything.

When the stress is ongoing, we may find ourselves in the third stage of the general adaptation syndrome: the *exhaustion stage*. The emergency has gone on long enough that we use up our energy reserves. We have been trying and failing to recover from the emergency, and we're now exhausted. We feel tired and possibly depressed, anxious, and unable to handle the loss.

If we stay in this stage long enough, our immune system starts to fail, leaving us vulnerable to brand-new emergencies. There is a well-documented link between long-term stress and increased rates of illness. Many bereaved individuals find themselves more susceptible to flu and colds and even to more serious disease, such as cancer. In a review of the research, Margaret Stroebe, Henk Schut, and Wolfgang Stroebe report that the death of a loved one (in particular the death of a spouse) increases the mortality risk of the survivors from many different causes. In this case, the survivor is said to have died of a broken heart.[6]

In all of these stages, we're still reevaluating our lives, we're still trying to adapt to the change that has been forced upon us, and we may be successful in our efforts to do so. However, if we cannot find a way to manage the stress of loss, the physical exhaustion we face leaves us at increased risk of disease and illness.

There is no definitive amount of time these stages last, no guarantee that they will be resolved after a set period of suffering has passed. We all grieve in our own way and in our own time. Most bereaved find that they eventually reach accommodation with the loss, inasmuch as that can be thought of as recovery. Most do so without requiring medical or clinical intervention.

My memories of the alarm stage remain murky. I have vivid but disjointed memories (sometimes called *flashbulb memories*—memories of the moments that surround emotionally evocative and important events) of some of the evening Chris died, but no real sense of the way things flowed, one thing to another. In the brain, cortisol affects the amygdala, hippocampus, and prefrontal cortex. These three brain regions are involved in our emotional responses, in the creation and storage of memories, and in what is known as "working memory." Working memory is the information we pull from storage and hold "online" for use in problem-solving. Researchers have shown that both good and bad memories that have a strong emotional component are remembered better than are memories of routine, everyday, and relatively unemotional events. The former are flashbulb memories, and they are the ones I have of that evening.

FIGURE 3.3. My favorite picture of Chris and his beautiful smile. He was just about to get out of one of his many stays at the hospital. He was happy.

I remember the empty, blank look in his still-open eyes. I remember feeling as though I couldn't get a breath and noting, almost dispassionately, that my hands were shaking as I reached for the phone to call 911. I remember trying to move him, to pick him up, and failing miserably. I remember seeing him on the living-room floor with the paramedic trying to restart his heart, but I have absolutely no memory of him being moved from our bedroom to the living room. It was as though he just appeared there, hooked up to machines, the paramedic crouched over him. I remember standing in the driveway, my friends and neighbors around me, but I cannot for the life of me tell you who was there or how long I had been standing there. I remember asking where the ambulance was; I didn't see it. How would they get him to the hospital? And I remember someone patting my arm and telling me that the ambulance had already come and taken him, right in front of me but without my awareness of it at all.

I tracked resistance and exhaustion in my journal as I struggled to adjust to the abrupt changes his absence created. There are numerous entries in my journal about not being able to sleep, about missing him and hopelessly, helplessly wishing he could come home. About being lonely but also about the benefits of solitude. About reconnecting to the world and my hesitations and fears in doing so. About the painful truth that comes with recognizing the dangers of reconnecting. I'm not sure I can do this again. I'm not sure I can watch someone else I love die.

4

THE EMOTIONS OF GRIEF

THE FIRST SIX MONTHS

I don't remember crying on the day Chris died; I think I was in shock. In the days that followed, I often felt as if I could not stop crying. My sister told me recently that for the first year after he died, I was crying every time she called me.

From the emergency room at the hospital, I faced the awful task of calling family and letting them know what had happened. Making those calls to his children and trying to help them with the shock and tears over the phone was one of the hardest things I've ever had to do. I know that I also called my siblings, but I don't remember what I actually said in any of these calls. My sisters (bless them) dropped everything and came to help. They gave me something to help me sleep, they made sure I ate, they cleaned my house and started the process of gathering up his things and going through the paperwork we all will leave behind. I deeply appreciated all of it—I needed the help. I think I sat on the couch in the living room, wrapped in a blanket (I couldn't seem to get warm), weeping.

They left after two weeks, first one, then the other, and I had gotten to the point where their leaving was a good thing. I'm sure they had reached whatever the limit is for watching someone you love deal with pain and sorrow. I remember thinking, "Please, stop asking me if throwing away a ten-year-old water bill is OK. He is a packrat and keeps everything, so it needs to stay here. Yes, I want to keep his books, even the ones in French (which I don't speak or read)—he loves books, he'll want them here. Please don't put his favorite sneakers in the 'going to Goodwill' box. I know I can't wear them, but I just can't give them away. He will need them."

Note the use of the present tense in these thoughts. There was a great deal of denial in my reluctance to let his books and shoes and even the detritus of his paperwork go. I knew that it was silly to hang on to these things that he would never need again, but that did not stop me from wanting to do so. It took a while to get to the point of not snapping my head around toward the door if I heard someone on the porch or out on the street. It took a while to stop the sudden thought, "Oh, thank goodness. He's finally home." I have since read that this kind of denial is relatively common at the beginning of the grieving process. It was gut-level denial, deep-seated and implacable. I understood that he was never again going to walk through the front door, but I just could not face that fact.

The change in the house when my sisters left was immense. The first time I sat down in the living room on my first evening truly alone, I realized that the empty house was so still it actually had a hum. I think it is an A.

One day about a month or so after Chris died, I was coming back to my office when I noticed a student of mine in my colleague's office next door. The student called out, saying she was talking to all her professors and needed to talk to me as well. She wanted to explain why she had been absent recently. I did not say that I hadn't noticed, but, locked in my own grief, I had not. She told me that her boyfriend had passed away. My first thought was, "Oh no, he's way too young!" She wanted me to know that his passing was the cause of her recent academic difficulties. She was having trouble finding a reason to get up in the morning. When she said that, I lost it. I found myself sitting in someone else's office, weeping, surrounded by people I don't really know all that well, struggling to pull my emotions back, with (scarily) little success. If I'd been asked prior to Chris's death, I would have sworn that I would never cry in public, and yet here I was, doing just that. I felt as though I were spinning out of control.

The intensity of these bouts of crying diminished as time passed. What had at the beginning been full-out wailing gradually ebbed to silent tears that lasted sometimes only for moments and then eventually to a brief filling of my eyes. I made a promise to myself that I would find a way to get through one whole day without even this very subdued form of crying. I finally made it on the 366th day after he died. When I realized I had achieved my goal, I promptly burst into tears—I counted it a success anyway.

Worse than crying—in fact, worst of all, at least for me—were the sadness and its partner in grief: the fruitless yearning for him. To yearn for something is to wish for it, to desire it strongly. The thing we're yearning for is often something that we cannot have; that is certainly the case in grieving. The bereaved want, above almost everything else, for the lost person to be returned to them. But that is impossible.

Sadness and yearning are aspects of grieving that, for me, are incredibly persistent. Crying eventually tapers off, wandering and searching decrease, but yearning for the lost person—that may well be permanent. As I write this, it has been four and a half years since Chris died, and I still miss him every single day and wish he could be here.

This yearning has changed with time, just as the crying changed. It isn't the sharp stab of pain that it used to be. Now it is more of a brief wave of sadness, an ache when I see his picture or think of things we used to do or go by our favorite restaurant. On some days, the feeling of yearning is stronger than it is on others. On some days, it consists of the brief thought "I miss you," and then I go on with the day. I suppose I've gotten used to missing him. I've adapted to his absence. It's just become a part of who I am now. He's not here, and I miss him. I know he cannot and will not return, but that does not stop the yearning and the sad ache.

THE FUNCTION OF EMOTION

If you have felt sorrow or pain or rage—which, without doubt, everyone has at some point in life—then you might be wondering why these emotions that can be so hard to bear are still with us as a species. Why do they remain part of our makeup as human beings? Wouldn't life be easier without them?

All the emotions we feel, whether happy or sad, are part of the stockpile of reactions available for us to use when we need to respond to something in the environment, solve a problem, or decide what to do next. Robert Levenson, a professor of psychology at the University of California at Berkeley, says that emotions serve several vital functions. They organize the responses that we make to an event in the world, including our facial expressions, our tone of voice, our muscle tone, even our endocrine (hormonal) response. Emotions also activate some behaviors and inhibit others. Emotions act to communicate what we're feeling to other people, pushing us toward some things and away from others.[1]

It isn't hard to think of a function for positive emotions. Emotions such as happiness, especially when accompanied by a happy facial expression, can elicit a shared smile from someone else, help reduce aggression and tension in potentially dangerous situations, and help us bond with other people. And they feel great! Negative emotions such as sadness and anxiety and anger also serve a purpose, motivating us to move away from someone or something and communicating that we need help or support from others.

SADNESS

About six months after Chris died, I had a revelation as I met up with a friend for a drink after work. As with so many of my friends, her conversation started off with "How are you doing?"—a question I appreciated and dreaded in equal measure. A line from a song used to run through my head when people asked me this: "I'd say I'm ok, but I'd be lying."[2] She gave me a long look. "Do you think you need antidepressants?" "Lord, no—I don't want to take drugs. I don't think I'm depressed." Although antidepressants are very effective for depressed individuals, they are not helpful for someone who is not actually depressed (in fact, they can be harmful). There was a perfectly valid reason why I was only just OK. My best friend, my lover, my husband, my partner of thirty-six years was gone.

My friend looked at me for a moment longer. "I don't think you're depressed," she said. "I think you're sad." And I thought, "That's exactly it. I'm profoundly sad."

There is general (although not total) agreement among the experts that sadness and depression differ. Sadness can look very much like depression, but there are moments when the sadness lifts, and you can laugh at something or feel comforted by the words of a friend. Grieving individuals often have moments when they smile or laugh, even amid deep sorrow. These moments of absence, almost of remission, are typically not present in depression. Sadness usually lasts for a shorter period than depression, and, best of all, sadness goes away—never as quickly as one wants, but it does remit. Depression usually does not go away, at least not without help. Sadness is also not generally accompanied by feelings of worthlessness, which is an unfortunate characteristic of depression. And, most importantly, sadness has a clear and usually easily identifiable trigger. Sadness is *about* something, caused by something that can easily be pointed to as the cause. Depression does not have an easily identifiable source. [3]

Sadness, even the deep sorrow of grief, also serves a purpose. George Bonanno, a professor of psychology at Columbia University, says that one benefit of grief is that it "gives us a job" to do. Grief and the sadness that comes with it demand that we "slow down, . . . turn inward," and reevaluate what is important to us and how we're going to go on without the person we've lost.[4]

Sadness is evoked by the perception of a loss that we cannot change. We feel sad when we lose anything that is or was important to us—prestige, money, social status, a relationship, or an individual person. Evolutionary psychologists say that sadness triggered by loss serves to help us recalibrate our thinking.

Research supports the idea that sadness can be beneficial. For example, Joseph Forgas, a social psychologist at the University of New South Wales in Australia, says that negative moods, such as sadness, have distinct benefits in particular situations. His research has shown that sadness can change *what* we think about (it's easier to remember sad events in our lives when we're already sad and happy events when we're happy) as well as *how* we think, making us better "bottom-up" thinkers. Bottom-up processing focuses on the immediate environment around us, and when we engage in this kind of thinking, we pay more attention to details in the environment, which gives us better memory for those details.[5] For example, suppose you walk into a friend's house and smell the wonderful scent of chocolate. You perceive (understand) that someone in the house is making something chocolatey and hope that it is for dessert. You don't have to rely on any other information to understand that there is chocolate nearby. All you need is the most basic immediate sensory information—the lovely aroma of chocolate. No prior knowledge is necessary.

Several studies have found that the valence (positive or negative) of our moods has a strong effect on our thinking. Emotions with a positive valence include joy and happiness, whereas anger and sadness have a negative valence. The valence of the emotional response we have to an event may act as a signal about what information-processing strategy we need to use in that particular situation, and it thus influences not only what we remember but also how we use that memory in problem solving.

In some situations, our thinking may tend to benefit from a more positive mood. In a positive mood, we rely more on what we already know, our memory, as well as on heuristics. Heuristics are generalizations, rules of thumb, or mental shortcuts that make solving a problem faster, but they also increase the chances of an inaccurate or irrational conclusion. Stereotyping is an example of a well-known heuristic, as is the "halo effect," or the tendency to let a positive or negative first impression of a person color everything else that person says or does.

One of the more interesting situations in which we can see the effect of emotional valence on thinking is when we're asked to make judgments about other people in social settings—something we do routinely. Galen Bodenhausen, Lori Sheppard, and Geoffrey Kramer discovered that not all negative emotions are equal in the way they influence social appraisal. They hypothesized that because anger, which has a negative valence, is usually associated with immediate threat and more often requires fast action, angry people tend to act impulsively and to rely more on heuristics or what they already know.[6]

Sadness, which also has a negative valence, is triggered by long-term problems with the circumstances of our lives. Fast action is not the best approach to these sadness-triggering events. Slower, more thoughtful consideration of the problem will result in a better solution. One benefit of sadness may thus be that when we're sad, we approach solving problems differently than we do in the heat of anger. When we're sad, we pay more attention to details, to bottom-up processing, we're less affected by heuristics, and we're slower to react in general.

Sadness also tends to increase our motivation to persist at a difficult task. What is more difficult than reevaluating and revising your entire life? Sadness can also increase generosity, perhaps making it easier for the grieving person to reconnect with others after loss.[7]

CRYING

If you ask people what separates human beings from other animals, including other primates, the answer you will usually get is likely to be that only humans use language, as distinct from communication. Animals communicate with one another and with us, but language use is uniquely human, and it is extremely effective in getting other humans to come to our aid. Language use does separate humans from other animals, even other primates, but it is not the only characteristic that does so.

Humans are the only animal to shed tears as an adult because of an emotional trigger. There are different kinds of tears, beautifully described by Ad Vingerhoets in his book *Why Only Humans Cry* (2013).[8] There are *basal* tears that we constantly shed with each blink of the eye, preventing the eye from drying out and coating the eye with a protective lubrication. Animals, including humans, also shed *reflexive* tears when their eyes are irritated (as when we're slicing onions or when the wind blows dust into our eyes). These tears are protective and are shed to get the irritant out of the eye. But only humans shed emotional tears in response to overwhelming and strong emotions, both positive and negative in valence. I've heard it argued that elephants and gorillas shed emotional tears, but the research does not support this contention. Elephants apparently lack tear ducts and so cannot cry, at least in the same way that humans do.[9] Gorillas (primates like us) do vocalize when they're in distress but do not seem to shed tears in response to emotions, as we do. It is also quite difficult to tell if gorillas or elephants are feeling a particular emotion because they do not have a way to tell us.

We need to make a distinction between *crying out* (sending an auditory signal of distress) and *crying tears* (a visual signal). Both humans and other animals cry out as infants to let others know that they are in distress and need help. That human vocal signal is loud, insistent, and difficult to ignore for good reason. It is an obvious signal that something is wrong. Animals use specific distress calls to signal that they are in trouble, while humans use both sound and tears to signal their pain or distress and sometimes their happiness.

The question for researchers has been why crying persists in humans past the point of language acquisition. Telling someone that we're in pain, using language to express precisely what the difficulty might be, is a much more effective signal than are tears, yet crying persists across our lifespan. Why do we do this?

There are several explanations for why we humans cry. One of the oldest theories is that crying resulted in restoration of balance in the body. For example, in 1981 William Frey and colleagues suggested that crying releases toxins that had built up in our bodies and that we need to get rid of. Without crying as a release, these toxins would accumulate to the point where they could cause disease or dysfunction, maybe even death.[10] This idea is closely related to the similar proposal that crying is cathartic, that it releases negative emotions that otherwise would build up to the point that they may induce significantly more negative emotional states, such as depression.[11] And then there is the notion that a good cry will reestablish homeostasis or balance in our physiology—a sort of "it feels better when I stop crying" situation.[12]

Anecdotally, the restoration-of-balance ideas are very popular. People will often say that they feel better after a good cry, so crying must lessen the painful emotions that generated it. In the same vein, many people will tell you that shedding tears releases some sort of toxic buildup in the body or mind. However, in a careful study in 1994 James Gross, Barbara Fredrickson, and Robert Levenson found that none of these popular explanations for crying was supported by the data.[13]

After using the tried and trusted "watch a sad movie" technique to evoke sadness and tears in their participants, these researchers found that people were typically even sadder after crying than they were before the tears started to fall. However, the research examining whether a good cry can be cathartic may not be quite so cut and dried. The social context in which the crying occurred might be biasing the results of the studies. When people were made to cry in the lab, where it can be assumed that they knew they were being observed by dispassionate researchers, a good cry did not improve mood or make the crier feel better. In fact, most participants in lab studies of the effects of crying reported that they felt worse after crying.

However, if the crying took place naturally, outside of the laboratory and in a very different social setting, a good cry did seem to produce emotional benefits for the crier.[14] It is much harder to study naturally occurring crying than it is to study crying in the lab. One major problem is that natural (not lab-induced) crying is hard to observe (the researcher has to be at exactly the right place at the right time to witness it) and harder still to elicit (there are some potentially serious ethical problems involved in making someone cry about something real, not fiction). And, by definition, natural behaviors happen in natural settings, outside of the lab. They happen without the control the experimenter would normally have in the lab, allowing "extraneous" variables to influence the behavior under study (such as whether we like the people we're with or how many people are witness to us crying or what our culture tells us is appropriate when we feel like crying). This makes natural crying much more likely to be influenced by variables other than our emotional state at the time and so much harder to assess. The best that can be said is that the laboratory research has yet to find good evidence that crying produces catharsis.

Chemical analysis of tears also showed that there is nothing toxic in the chemical makeup of the tears we shed. In addition, rather than reducing activity in the sympathetic nervous system so that homeostasis can be restored, crying increases this activity. Logically, this increase should then throw the systems in the body even more out of balance than they were prior to the shedding of tears. In a review of the literature in 2008, researchers were unable to find consistent evidence that crying is beneficial to one's physical or emotional health.[15]

If the reason for adult crying doesn't have effects on our emotional or physical health, then perhaps it is related to some other aspect of our psychology. One theory suggests that crying might serve as motivation for us to fix the problem because crying produces an aversive state of high arousal that we find unpleasant. This unpleasant feeling might push us to solve the problem that made us cry to begin with.

Another theory suggests that crying may be a signal to other humans near us that we're in trouble and need help. Vingerhoets suggests that crying in adults as well as in children is an attachment behavior that evolved to solicit help from others and to keep them nearby. For example, Vingerhoets and colleagues asked participants to read short stories describing crying and noncrying people. These volunteers were then asked how they would judge the person described in the story, how the situation described in the story would make them feel, and what their response to the person would be—would they help the person or not? Although their participants reported that the crying person made them feel uncomfortable and that they tended to judge the crying person as more emotional than a noncrying individual, they also reported that they would be much more likely to offer help to the crying person than they would to the noncryer. Participants also reported that the crying person made them feel less anger toward them, which may support a related function of crying—to reduce aggression toward the crying person.[16]

EMOTIONS AND THE BRAIN

All of our emotional responses to events in the world are the result of brain function. However, no one brain region is responsible for any single emotion that we experience—no "happiness" center or "sorrow-inducing nucleus." Our emotional responses are instead the result of the systems I discussed in chapter 1, not just of the limbic or emotion network alone. Bonanno says that the way emotions feel is the result of all of the networks in the brain working together and of "the ways we interpret the circumstances we find ourselves in."[17] Interpreting what's happening around us and coming up with a plan to deal with whatever it is constitute a whole-brain activity.

The limbic (emotional) circuit in the brain is, not surprisingly, involved in our emotional responses. Several parts of the limbic circuitry stand out. The frontal lobe, in particular the orbitofrontal cortex, the nearby anterior and posterior cingulate cortexes (ACC and PCC), and the amygdala are major players in our emotions.

Think of the brain as shaped roughly like your hand closed in a fist. Your thumb, wrapped around the rest of your fingers is analogous to the temporal lobe—that is, the cortex at your temples. If you peeled your thumb up and away, you would see the amygdala resting toward the center of the brain. The ACC and PCC are deep layers of cortex, underneath the outermost layers on the midline of the brain, with the ACC toward the front of your head, and the PCC toward the back, as indicated by their names. The orbitofrontal cortex is where you think it would be: at the front of the brain, above the bony orbits surrounding the eyes. The lateral view in the top half of figure 4.1 illustrates what you might see if you were standing to one side outside of the brain. The medial view in the bottom half of the figure shows what you would see if you could stand in the middle of the brain, between the two hemispheres, looking at just one side of the brain.

Recent research has suggested that happiness, sadness, and fear might rely on the same neural circuitry and just require the

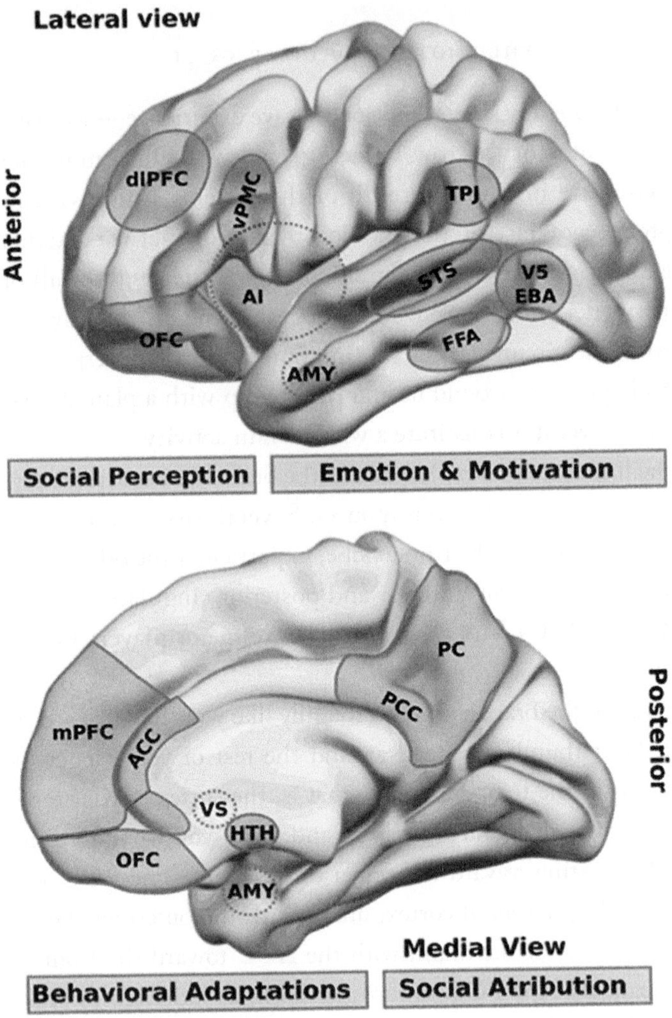

FIGURE 4.1. Emotion centers in the brain, in particular the orbito frontal cortex, the anterior and posterior cingulate cortexes, and the amygdala. *Abbreviations*: ACC = anterior cingulate cortex, AI = anterior insula, AMY = amygdala, dlPFC = dorsolateral prefrontal cortex, EBA = extrastriate body area, FFA = fusiform face area, HTH = hypothalamus, mPFC = middle of prefrontal cortex, OFC = orbitofrontal cortex, PC = parietal cortex, PCC = posterior cingulate cortex, STS = superior temporal sulcus, v5 = visual region of cortex, VS = ventral striatum, TPJ = temporal parietal junction, vPMC = ventral premotor cortex.

Source: Image from Wikimedia Commons, https://commons.wikimedia.org/wiki/File:Brain_areas_that_participate_in_social_processing.jpg, accessed March 13, 2025.

use of different subregions within each network. For example, two different regions within the orbitofrontal cortex might be responsible for coding the valence of the emotion, the relative pleasantness or unpleasantness of that feeling. In turn, different subdivisions within the amygdala might be coding for the arousal that those emotions produce. Fear is often thought of as a combination of unpleasantness and high arousal, but sometimes fear can be arousing but pleasant. Imagine the fear evoked by a potential attack in the dark versus the fear evoked by riding a roller coaster. Both activate our arousal system, but the first situation is usually seen as unpleasant, whereas people actively seek out the second.[18] The way we interpret what is happening around us feeds into the emotions that we feel.

The emotions we feel when we grieve undergo this same interpretation. The sadness we feel in the moment of loss might be interpreted quite differently than the sadness of recalling a memory of time spent with that beloved person.

There is even a circuit for guilt in the brain. A group of researchers at Duke University and Harvard University asked people in an fMRI scanner to read stories describing situations where someone's actions resulted in harmful consequences to others or harm to themselves or harm to no one (the latter a control "neutral" story). They found that the intensity of the feelings of guilt the participants experienced was reflected in the level of activity in regions of the frontal lobe, in particular the dorsal medial prefrontal cortex, toward the top middle of the brain map in figure 4.1. People also felt stronger guilt when someone else might come to harm compared to the actor harming themselves or producing no harm. And, finally, feelings of guilt evoked by stories of harm coming to another person activated a different circuit than did stories when someone's actions resulted in self-harm. This tends to support the idea that guilt, like all emotions, serves a purpose. Guilt helps us control our behavior in social situations—greasing the social wheels, so to speak.

We are a social animal. That simple statement means that we depend on other human beings and our interactions with them for our continued survival and quality of life. The old saying that "no man is an island" has been practically and scientifically found to be true. Human beings do not survive well in isolation. Our emotional responses evolved to help us interact with other members of our group. They are equally vital to our long-term health and survival.

THE CONSTELLATION OF EMOTIONS IN GRIEF

In those first months after Chris died, I would sometimes notice myself feeling happy, perhaps smiling at a joke or laughing out loud, and inevitably I would almost immediately feel guilty for having smiled or laughed, even though I knew that feeling guilty about being happy was irrational. Guilt is quite common in the bereaved. There is even a paper-and-pencil test you can take to measure feelings of guilt in bereavement, if you are so inclined, called the Bereavement Guilt Scale.[19] Researchers have identified a number of different causes for the guilt we feel, ranging from survivor guilt to guilt arising from a fear that something we did or failed to do caused the death of the person to guilt from feeling that we are either not coping well with the death (called "grief guilt") or are coping too well (recovery guilt).[20]

I know I felt guilty that I did not recognize that particular Thursday as Chris's last and that I was not with him when the end actually came. I also recognize that I had no way to know that he was going to die on that particular day and that my leaving him alone was an attempt to allow him to get some much-needed rest rather than abandonment. I know that feeling guilt was irrational, but it was there nonetheless.

Although sadness is generally thought of as a defining characteristic of grief, sadness is not the same as grief. When we grieve, we experience all the emotions that are possible for human beings. Researchers describe grief as a "complex of emotions, rather than as a single emotion"; it consists of sadness, anger, guilt, joy, anxiety, relief, confusion, frustration, fear, hope, yearning, and envy, just to name a few emotions that grieving individuals describe as part of their experience.[21]

None of these emotions lasts forever, even though they sometimes feel as though they do or will. Emotions are by definition relatively short-lived and ephemeral. Researchers have found that emotions such as sadness may last only a few seconds, whereas grief, an entire constellation of emotions, can last months and years. Emotions, brief though they may be, tend to happen in waves, with peaks and valleys in their intensity. The emotions we feel when we grieve have been likened to a stress reaction, and, explains Bonanno, "like any stress reaction, [they are] not uniform or static. Relentless grief would be overwhelming. Grief is tolerable, actually, only because it comes and goes in a kind of oscillation."[22]

The same can be said about the other emotions people experience. Happiness, anger, fear—all would be intolerable if they were static and unchanging. Imagine feeling nothing but anger or even nothing but happiness all the time. Not only would the person experiencing the unending emotion find it difficult to deal with, but so would family and friends.

Many researchers have noticed that even the recently bereaved, who would be expected to be feeling the emotions of grief more strongly than would people several years "out" from the loss, expressed happiness, even laughing when they were asked to recall their absent loved one. It may be that the ability to express positive emotions such as laughter and happiness even amid the pain of grief predicts better overall coping with the loss. I discuss this correlation in chapter 7 when I talk about resilience and coping.

The emotions we feel when we grieve are as individual as the people who experience them. We all had unique relationships with the lost person, and our emotions as we adjust and adapt to their loss are correspondingly unique. Grief is more than just sadness.[23]

5

PAIN

Nothing begins, and nothing ends,
That is not paid with moan,
For we are born in other's pain,
And perish in our own.
—Francis Thompson, "Daisy"

THE FIRST YEAR

Many of the entries in my journal in the year after Chris died feature the pain of grieving his loss. The pain changed over time, sometimes localized to a specific place, sometimes a more general, whole-body, empty, and aching pain that could send me to my knees.

Seventy-five days after he died, I wrote about the last few months of his life and our increasingly crowded calendar, now filled not with parties or trips to see the kids but with visits to doctors and appointments for tests. Stressful and difficult, these visits marked the end of his life. I wish they hadn't. I found myself lamenting the time I'd wasted during those months. All the times I could have told him that I loved him and didn't. The times I could have chosen to laugh something off but got pissy instead. Every fault, every missed opportunity, everything I wished I'd done differently. They felt as though they were sitting on my heart, crushingly heavy, making it difficult to breathe.

We had a memorial service almost two months to the day after he died. My family, some of our friends, and his children and grandchildren came to see his ashes buried in Savannah and to have the world's mildest wake afterward. I wrote about standing by the gravesite, my arms wrapped around my stomach, bent over from the aching, hollowed-out feeling, my sisters standing next to me in case I started to sink. My throat closed up, and I couldn't speak, so my younger sister (who is made of titanium, the strongest person I know) read what I had written to say about him. I just couldn't make words come out.

One hundred days after he died, I wrote about a former student who reintroduced herself at an end-of-the-year reception for current majors and alumnae. She said that she was now teaching statistics and that her university had given faculty free copies of my introductory statistics textbook to consider for use. She thought to herself that the author couldn't be the same person who was her statistics professor at school but then googled me, finding that it was. She told me she was adopting my textbook for her class and would be using it at her new university when she moved there to continue working on her PhD. I was overjoyed when she told me and looked forward to telling Chris all about it, but then I abruptly remembered that I couldn't tell him and had to turn away quickly because I was crying. The stab of pain doubled me over.

One hundred and fifty-eight days after his death, I went to a family wedding and broke down at the reception when my sister-in-law's parents, both then in their nineties, stepped up to the table to convey their condolences about his death. I sat there looking at two people obviously in love with each other still after many decades of marriage and burst into tears. The pain of envy and jealousy, regret and loss burned, and I was ashamed.

Three hundred and twenty-four days after he died, I wrote about looking through old photographs, thinking about which ones to send to the kids, and feeling as though I were holding my hand over a flame with no way to move it or put out the flame. Was this a test? How much pain can I stand? Fires, flame, and burning seem to be my metaphors for the experience at this point.

PAIN

Talking about pain is almost always done using metaphors. Pain is quite difficult to describe because it is quite literally "all in your head." What is painful agony to one person is nothing but a scratch to someone else. Debates about who hurts more are pointless because pain is completely subjective. Doctors often use images to help. If you remember visiting the pediatrician when you were a child (or if you've taken a child for a visit since then), you may also recall a poster on the wall with what look like emojis showing facial expressions ranging from a big smile to a very unhappy person crying. Called the Wong-Baker FACES Pain Rating Scale after the two child-development specialists who created it, this famous scale is now used worldwide to help children talk about their experience of pain (figure 5.1). Children are asked to indicate which face (and which number listed beneath it) matches the severity of the pain they are experiencing.[1]

Adults are sometimes asked to use the FACES scale, but because adults have more highly developed language skills than children, they're usually asked to simply use language to describe the pain they're feeling. I say "simply," but doing this can be

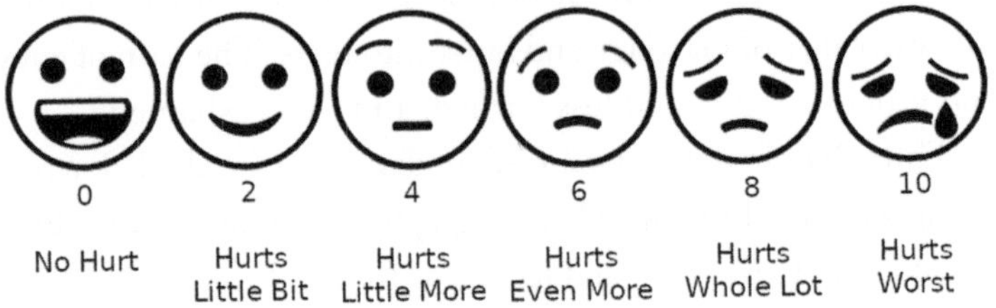

FIGURE 5.1. The Wong-Baker FACES Pain Rating Scale.

Source: Image from Wikimedia Commons, https://commons.wikimedia.org/wiki/File:Wong-Baker_scale_with_emoji.png, accessed March 14, 2025.

difficult. I vividly remember a discussion with Chris about describing pain. He thought that since pain is so very subjective, all the adjectives used to describe it didn't apply, and he wanted to just say, "It hurts, a lot." His doctors kept asking him to be more descriptive, making him "harrumph."

To help patients describe this difficult and subjective experience, doctors can use tools such as the McGill Pain Questionnaire, which is essentially a list of adjectives that can be applied to the experience of pain. The McGill Questionnaire lists seventy-eight words and asks the people taking it to select the words that describe the pain itself (e.g., *sharp*, *tingling*, *heavy*, *hot*), the way the pain makes them feel (*tiring*, *punishing*, *wretched*), and the evaluative aspect of pain (*annoying*, *unbearable*, *nagging*). The participants select the words that apply to their individual experience. Each word has a point value associated with it, and the points are summed to provide a score that ranges between 0 and 78. The higher the score, the more severe the pain.

These two pain-rating scales (there are others) were originally intended to evaluate physical pain. This is probably because Western medicine has a history of ignoring psychological pain, even suggesting that it does not actually exist.[2] Recently, however, a shift has begun in the way pain is viewed in the lab and in the clinic. Scientists and medical doctors have known for a very long time that laypersons readily accept the existence of nonphysical pain, even using the same words to talk about the psychological pain of a broken heart that they use about the physical pain of, for example, a broken arm. So, just how is pain defined, created, and perceived?

PHYSICAL AND PSYCHOLOGICAL PAIN

Researchers describe four stages of a painful experience. First, there are the physical aspects of pain, where on the body it occurs and its intensity. Then there is an emotional reaction to the pain. For example, we might feel afraid that it will happen again or that serious damage has occurred. Then there is a cognitive reaction to the experience, which consists of our expectations about the short- or long-term consequences of that pain, our previous history of painful experiences, and the context in which the pain happened. Last, we express (or sometimes don't express) that pain with facial expression, words, body posture, and so on.

Physical pain is caused by damage to the body or by inflammation. When we suffer a cut or a bruise, or sometimes even just when we experience something that *might* damage us (such as an extremely hot stovetop that we have not yet touched), specialized cells called *nociceptors* (*noci* is the Latin word for "to hurt") respond, sending a signal to the cortex of the brain. That signal sounds an alarm that something is wrong and immediate action is needed to fix it.

This signal travels to the brain via the spinal cord. The layout of the circuitry makes it possible to interrupt the signal enroute to the brain and, in doing so, to interrupt the experience of pain. Have you ever noticed that gently rubbing a small pain like a sting or a scratch can actually make the pain lessen or even go away? That is the take-home message of one of the most famous theories in psychological research: the *gate control theory of pain*. The researchers Ronald Melzack and Patrick Wall noted that light touch travels to the spinal cord faster than does the alarm pain signal from the nociceptors. Applying a gentle rub to the skin around some point of minor damage seems to "close a gate" in the spinal cord, blocking, at least temporarily, the pain signal heading for the brain. Of course, if the damage is severe, such as a broken bone, closing the gate isn't going to work, but for minor pains it is fairly effective.[3]

The brain also has a built-in system for reducing pain, called *analgesia*. The brain produces its own version of opiates called *endogenous opiates* (from the Greek word *endon*, meaning "originating from within the body") and *enkephalins* (from the Greek word *enkephalos*, meaning "brain," so coming from the brain). These chemicals are structurally quite similar to natural and synthetic opiates such as morphine and fentanyl and have an action on the cells of the brain that is similar to the opiates we're more familiar with. They effectively reduce physical pain.

Some very interesting recent research has found that synthetic opiates activate a structure within the brain cells that endogenous opiates do not, and this difference may account for the significantly higher risk of addiction seen with synthetic drugs compared to our naturally occurring analgesics.[4] Focus on these differences may lead to the development of less addictive pain-management drugs.

The cause of *psychological pain* (also called *social pain*) is the loss of a bond or attachment or membership in a group. Rejection or isolation from the group causes us pain because we are a social animal. We live in and thrive in groups. Isolation from the group is dangerous to our survival and can significantly shorten our lifespan. We need the group to survive. If you have ever been rejected by someone else or by another group, you know what social pain is. The pain of grief is caused by the loss of an essential bond with another human being, and to describe this kind of pain most people use the same words that they use to describe physical pain.

THE PURPOSE OF PAIN

Many people have wondered if we wouldn't be better off without pain in our lives. Very few people enjoy it, and even fewer seek it out, so why do we suffer this particular sensation? It's probably obvious why we have a system for feeling physical pain. Physical pain signals that damage to our bodies has occurred and we need to do something about it. The life expectancy of people who because of disease or genetic disorder cannot feel pain is significantly shortened. For example, Hansen's disease (a.k.a. leprosy) is caused by a bacterial infection that attacks the same nerves that carry touch and pain signals to the spinal cord. As a result, patients often lose the ability to feel pain and touch and can suffer serious long-term injury from even relatively minor tissue damage. They don't feel the pain, sometimes don't even know that they've hurt themselves, and so do not treat the injury. Life expectancy worldwide for people with this treatable disease is generally about five years shorter than for individuals without it.[5]

Psychological pain is generated by the loss of something important to us. That important thing can be another person, a relationship, our self-image, our dignity, or our hope for the future. Something or someone that we want to keep near us has been taken away, and it hurts.

It may well be that psychological pain is also serving as an alarm signal, in this case warning that there has been damage to a vitally important social relationship that could endanger our survival. Without other human beings in our lives, we don't last very long or very well.

Evidence for the effects of what is often called *long-term isolation stress* can be seen in the recent quarantine and social isolation produced by the pandemic. There was a significant uptick in psychological distress worldwide after the COVID outbreak began, showing up as post-traumatic stress disorder, confusion, anger, frustration, fear, and pain.[6]

Studies of the effects on grief and grieving individuals of the COVID pandemic and the social isolation put into place to combat its spread are ongoing. For example, Saachi Arora and Sangeeta Bhatia report that the experience of grief during the COVID-19 pandemic "led to heightened psychological symptoms of depression, anger, anxiety due to the lack of opportunity to bid a final goodbye to the deceased." These researchers note that rates of prolonged or complicated grief spiked during the pandemic, perhaps because of the restrictions and precautions placed on gathering together during the pandemic. These measures were very effective in slowing the spread of the disease but at the same time hampered the process of mourning the loss. These authors point out that "the process of mourning is an important protective factor against pathological grief." They also speculate that the isolation the pandemic necessitated led to high rates of *disenfranchised grief*, or grief that we cannot or will not express in public.[7]

Research has pointed to the benefits of the public expression of the pain of grief to both the bereaved and those less directly affected by the loss.[8] The bereaved can benefit from the public support offered and the reduction of the pain that has been generated by feeling alone in their suffering. The "public"—those not closely related to the loved one who has been lost—benefits from a reduction in their level of upset and feelings of helplessness that participating in the memorial service (or funeral or wake or however the loss is recognized) offers.

Christopher died at the beginning of 2018, before most of us even knew what the acronym COVID stood for and what a pandemic would look like. We were able to have a memorial service for him and to allow everyone to say goodbye, and I'm very grateful for that opportunity. A little more than a year after he died, the first wave of the virus hit, and abruptly, in the middle of a semester, my own and my colleagues' everyday interactions with students were yanked around 180 degrees. We went from in-person teaching to online teaching with only about a week to switch gears and learn how to do this. I never felt more isolated and alone in my life. Pictures on a computer screen quickly devolved into what I called "the wall" because most students did not turn on their cameras or just showed a still photograph of themselves, leaving me with a screen full of static names or faces, talking to no one and getting zero feedback. This is no way to teach.

In 2019, my longest-term friend (somewhere there's a picture of us, plopped down in a sandbox, playing next to each other, when we were both about a year old) passed away, and her memorial service had to wait because of COVID restrictions. When we finally were able to gather to remember her, a year had passed. For me, the delay highlighted the sense of distance I experienced between myself and her children and extended family. And in 2020, my mother, who was in assisted-living care at the end of her life, passed away with none of us able to be with her in her room at the end. Two of my siblings who lived nearby handled what visiting was possible, but I don't think Mom ever understood why they stood outside her room's window and waved and told her they loved her but wouldn't come in to say hello. The confusion she must have felt breaks my heart. Her memorial service was delayed almost a year so we all could travel back home to say goodbye. I don't know if the pain of her loss would have been lessened by an immediate memorial ceremony. It's hard to know what something that didn't happen would feel like. What did happen, the delayed memorial for her, felt odd, at least to me.

Research has suggested that delaying traditional mourning practices—having a memorial service for the departed, gathering with family to remember the lost individual, lifting a glass to toast the departed—can create the feeling that the loss was unjust, which can lead to increased feelings of anger and the urge to cast blame on someone or some institution in order to make sense of the loss. Being unable to mourn the lost person within a reasonable time because of pandemic restrictions may have made adjusting to the loss much harder to do.

ARE PHYSICAL AND SOCIAL PAIN THE SAME THING?

The observation that we tend to use the same language to talk about both broken limbs and broken hearts suggests that they might be one and the same thing. But they probably are not, or at least they are not exactly the same. For one thing, physical pain begins with tissue damage and a signal to a nociceptor. In psychological pain, there is no tissue damage, and the nociceptors are silent; no signal from them is going to the cortex.

Another important difference between the two has to do with how long each type of experience lasts. William James, one of the founders of psychology as a science in the United States and author of one of the first textbooks on psychology in 1890, noted that people consistently reported that they would prefer to experience physical pain than the social pain of loss or rejection. This is because physical pain ends, whereas psychological pain seemingly goes on forever.[9] This might well be another reason grief is described as never-ending.

In addition, the way physical and psychological pain interact with our memory systems seems to differ. It is possible to recall that you suffered physical pain from tissue damage, but it is not possible to relive that pain. For example, when I was four, I broke my arm. My best friend, Patty (yes, the same one I mentioned earlier), dared me to do something stupid, which of course I tried, and it ended with abrupt contact between a cement patio and my arm. I remember that it hurt, I remember screaming and crying all the way back to my house, Patty's mom trying to calm me down, looking almost as upset as I was. I remember the nuns at the hospital, who were scary looking, and the big, ugly, heavy, hot, and scratchy cast I had to wear for the rest of the summer, but I don't, thank heavens, remember and I cannot relive the actual pain itself. I think about that experience now, and all I recall is that it must have been painful, but none of the words that describe pain come to mind. I don't remember what the pain felt like. It makes evolutionary sense that we cannot remember physical pain, just the circumstances that created it so we can avoid it next time. No more dares from Patty.

In contrast, psychological pain can be and often is relived and reexperienced, relatively easily at that. If you've ever been rejected by someone you wanted to like you or approve of you or, worst of all, someone you wanted to love you, then you can probably recall what that felt like.

The research backs up this claim. For example, participants in one study were asked to write an account of either an experience involving physical pain or social pain, indicating how severe the pain was at the time it occurred using the FACES Scale, the McGill Pain Questionnaire, and another pain measurement tool called the Pain Rating Index. They also reported how long ago the painful experience happened. After writing down the story and rating the initial pain, the participants rated the amount and severity of any pain they were feeling at that moment during the experimental recall of the event. The participants' ratings of physical and social pain at the time they occurred were statistically equal. Their memories of how much the pain hurt when it initially happened were about the same for both kinds of pain. However, the amount of pain experienced at the time of recalling the event was significantly higher for the psychological pain than for the physical pain. It was much easier to recall psychological pain than it was physical pain, and that psychological pain was easily pulled out of memory and into the here and now.[10] This may be another reason why researchers say that grief lasts forever. That specific psychological pain is easily accessible, available with the appropriate trigger, for the rest of our days.

It is possible that the reasons for these differences between physical pain and social pain have to do with the way the brain experiences and processes pain. Researchers who study pain circuits in the brain focus on two component parts of the experience of pain, the *sensory* and the *emotional* or *affective*. The sensory component comes from the nociceptors, which tell the brain about the intensity and location of the painful stimulus. The emotional (or affective) component bypasses the nociceptors, stimulating the limbic system directly.[11] The affective component of pain has to do with how upsetting or distressing the experience of the pain is.

The sensory and affective components of pain can operate separately or jointly. It is possible, although unusual, to feel pain (the sensory component) but to have no emotional response to it, to know that it is there but simply not to care about it. For example, a condition called *pain asymbolia* results from damage to the insular region in the cortex (a part of the limbic circuit in the frontal lobe), so that the person who has this condition can feel the pain but also tells the doctor that the pain is not theirs, as if they disowned it. Because it isn't their pain, it doesn't bother them.[12]

The two components also interact. Being in social pain can make physical pain worse, and being in physical pain can make social pain more intense. Participants in a study at the University of California at Los Angeles were injected with a drug that temporarily induced an inflammation response, activating the sensory component of pain. While their immune systems responded to the drug, the participants were asked to assess the level of connection they felt with other people and their feelings of depression. Compared to participants who had received a placebo injection, those who experienced inflammation reported that they felt significantly more socially disconnected, isolated, and depressed.[13] Other studies have shown that having a low tolerance for physical pain is linked to a low tolerance for social rejection or social pain.[14]

The medications we use to alleviate pain—both the over-the-counter medications that we often take to reduce minor physical pain, such as acetaminophen, and the drugs available by prescription only, such as morphine—also significantly reduce both physical and emotional or psychological pain.[15] This may be because of the way these drugs affect how the brain responds to a pain signal, regardless of whether that signal originated in physical damage or in the loss of an important social relationship.

PAIN CIRCUITRY IN THE BRAIN

Multiple streams of evidence point to a strong similarity and connection between the social pain of rejection by a group or loss of an important relationship and the physical pain that results from damage to our bodies. The same language is used to talk about both kinds of pain, the same kinds of analgesics mitigate both types, and the two types interact with one another, amplifying or reducing our experience of them. Thus, it might not be a surprise that researchers have found that both types of pain are processed, moderated, and controlled by just one circuit in the brain, albeit a complicated one.

The sensory and affective components of pain are thought to be handled by two related parts of a single circuit in the brain. The dorsal (or top) parts of the ACC and the anterior, or frontmost, section of the insula (the anterior insula), described in chapter 2, contribute primarily to the affective component of pain—how unpleasant we find the pain to be. And because there is often no sensory component to social pain, researchers also think that these two cortical regions are particularly important in coding the pain of loss and grief.

Other cortical regions code for information in the sensory component of pain—that is, the location and intensity of the pain. Two regions, the *somatosensory cortex* in the parietal lobe and the *posterior insula*, code for the sensory aspects of pain. Figure 5.2 illustrates these components of the pain circuitry in the brain. The insula, buried deep inside a fold in the layers of cortex, can be seen, along with the ACC.

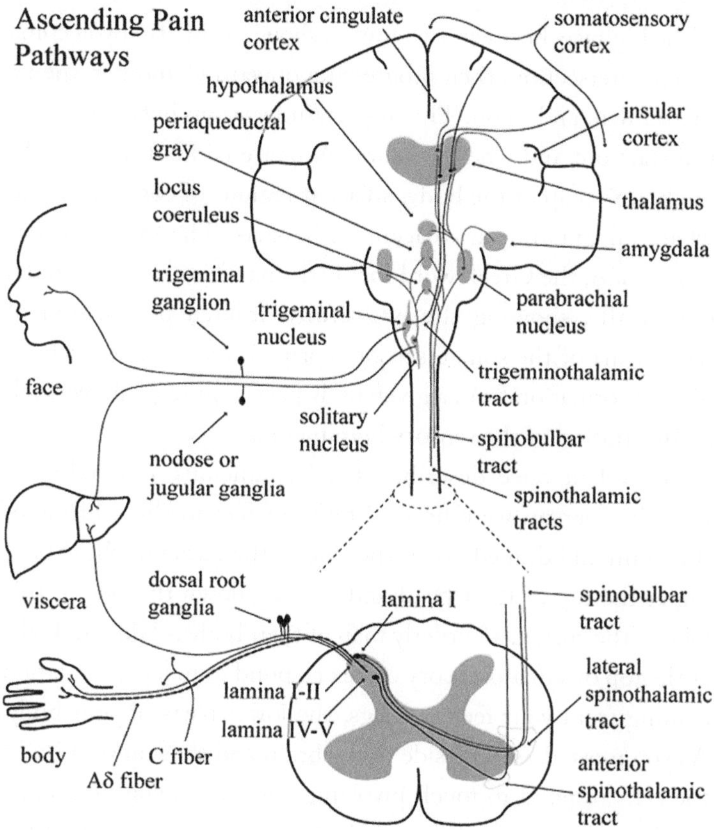

FIGURE 5.2. The pain circuitry in the brain.

Source: Image from Wikimedia Commons, https://commons.wikimedia.org/wiki/File:Ascending_pain_pathways.tif, accessed March 14, 2025.

The somatosensory cortex is primarily responsible for processing information about touch, such as where on the body the touch comes from and the type of touch experienced (pressure, rubbing, friction, or pain, for example). The somatosensory cortex is described as having a “map” of the human body laid out across it. It's obviously not a literal map. Imagine instead a map of the highway between two towns, showing that first you come to one intersection, then progressing onward to another, then a small town, and so on. The important locations between where you start out and where you are going are laid out in order. In the brain's map of the body, adjacent regions of cortex code for what is happening in adjacent body parts. The map, called a *homunculus*, the Greek word for "little man," is often illustrated graphically, showing the beautifully ordered progression of which part of the somatosensory cortex is responsible for handling information from a given body part. Figure 5.3 shows such an illustration of the sensory homunculus.

A student once described the homunculus as looking as though a person were hooked by their feet to the midline of the brain and draped down the side of the parietal lobe. Starting at the top of their head and moving down the side of this lobe of the cortex, an orderly map of their body is laid out. Cells at the top of somatosensory cortex respond to touch stimulation coming from your feet and legs, the lower parts of your body. As you move down the side of the brain and thus up your body, the cells respond to touch involving your torso, your arms and hands, and your face. The relative size of the body part on the homunculus in figure 5.3 indicates that body part's relative sensitivity to touch (or pain). Lips and hands are very sensitive, and so a lot of the homunculus tissue is devoted to them. The backs of your knees are not very sensitive, and so not many homunculus cells respond to them. This organization scheme means that this region of cortex can tell exactly where on the body a touch or a painful stimulus originated.

The insula is also organized, but not in exactly the same way as the sensory cortex. Instead of being organized like a map of the body, cells in the insula are organized by the kind of

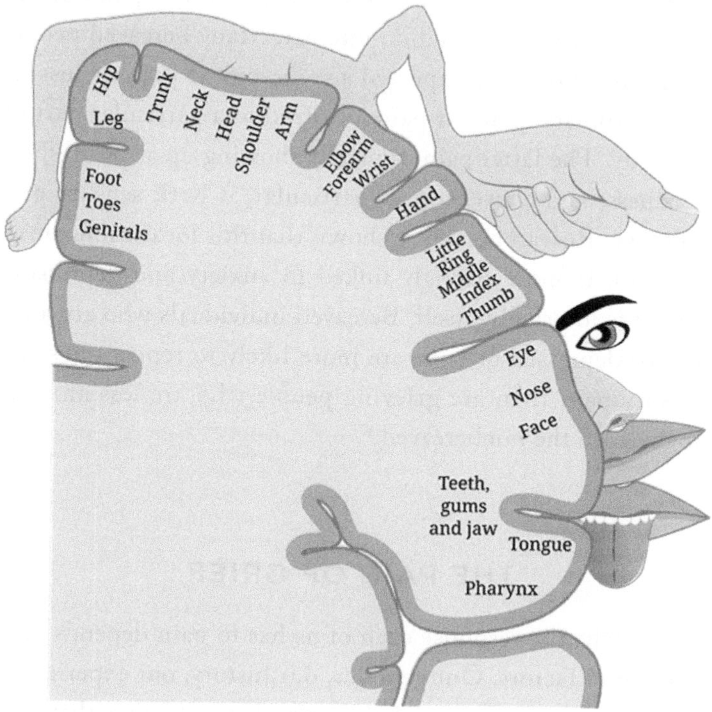

FIGURE 5.3. The sensory homunculus.

Source: Image from Wikimedia Commons, https://commons.wikimedia.org/wiki/File:Sensory_Homunculus-en.svg, accessed March 14, 2025.

information they work with and the job they do. Cells at the front of this brain region code for the subjective intensity of our experiences and our perceptions. Cells in the middle of the insula integrate the information sent to the insula from sensory and affective regions of the brain.[16] The posterior or back part of the insula is responsible for our awareness of the internal state of our bodies, helping us interpret our reactions to stimuli around us. The insula also codes for the sensory aspects of pain.[17]

Remember that the two component parts of pain, affective and sensory, can and do interact. Many bereaved people experience both the expected psychological pain of loss as well as somatic pain, or pain localized to a particular part of the body. The latter pain, typically showing up as headaches, dizziness, and chest pain in particular, is both sensory and affective. Researchers have shown that this localizable physical pain is more strongly linked to anxiety and depression than it is to grief by itself. Bereaved individuals who are anxious or depressed or both are more likely to report physical, sensory pain than are grieving people who are less anxious or than are the nonbereaved.[18]

THE PAIN OF GRIEF

Ultimately, the response each of us has to pain depends on a number of factors. Our genetics, our history, our experiences with pain in the past, and the immediate situation that surrounds the creation of that pain—all influence how the brain responds to pain and, in doing so, how we experience it.

The genes that create our sensitivity to pain have begun to be identified. They are the same genes that create *receptors* for the endogenous and exogenous opioids that the brain uses to process information about painful stimuli. Specific chemical messengers in the brain stick to specific receptors and command the cell the receptors are part of to send messages (or to stop sending messages) around the brain. There are naturally occurring differences in the genes that tell the body to make these receptors, which results in natural differences in the number and types of receptors each of us has. The result of this variability is that each of us differs in our individual sensitivity to pain and the effectiveness of drugs used to alleviate our pain. The variability in individual sensitivity to pain is also likely part of the reason why each of us experiences the pain of grief in our own unique way.

Genetic testing can predict sensitivity to both physical and social pain by identifying which version of these pain genes (different versions of genes are called *alleles*) we carry. Different alleles tell the body to make different numbers of receptors. Fewer receptors mean less of a response to the endogenous opiates carrying the pain signal, and vice versa. If you possess one version of the genes that create opioid receptors, you will have a heightened sensitivity to pain, both social and physical. If you have another version, you may have a diminished reaction to pain and to opioid drugs. This kind of data helps explain at least in part why the pain of grief varies from person to person. Each of us feels pain in our own way, and each of us grieves in our own way. I can't help but feel that the two events are linked.

6

GRIT, RESILIENCE, AND GRACE

Don't look back.
You're not going in that direction.
—Unknown

FOUR YEARS OUT AND COUNTING

So far in this book, I have been consulting my journal for the story of Chris's death and the ways I have experienced it. But with this chapter, I don't have that resource. My journal has been about the emotions, the pain, the loss of grief, and the questions I sought answers to. I have not, at least overtly, focused my journaling on the process of moving forward, of coping with that loss. I think this is because that process of adaptation happens slowly, in tiny increments, some so small that they go unnoticed even by the person experiencing them. But if I lift my head up and look back over the past several years, I can see that I have moved forward. I have been adapting.

I've made a number of changes to my life; some are huge, some quite minor. One major change is that I'm now (as of 2024) retired, in large part a decision made so that I could spend more time writing.

The other major change is that I sold the house Chris and I had purchased for our retirement. The initial plan was that I would retire, and we would sell the house we were then living in and move to the house nearer to the kids and grandkids. That's not how it worked out, though. Without him there, it was just a house that I would rattle around in, too big for just me and too crowded with memories. I was a bit surprised to discover that I just couldn't be there. Even though we hadn't fully moved in and lived there only in the summers when I wasn't teaching, there were too many memories. I think the problem was that this house was the one we had made so many plans for, and now those plans were abruptly moot. Clearing out the home we were going to retire to was difficult, much harder than I thought it would be because we hadn't lived there for long. It felt right to flip our initial plan and to stay in the house where he died because so much of my life, both with him and without him, had already been spent there. I still live there.

I made uncomfortable and difficult decisions about what to do with the things his death left behind. His guitars I gave to his kids with a prayer that they use them. The books in French (which I don't speak or read) I donated to a school in Haiti in his name. I still find golf tees, balls, and golfing paraphernalia in unexpected places. I had learned to play golf because he had taken it up again. I had quickly discovered that if I wanted to talk to him, I would need to learn to play because that's where he would be if he had any free time. For a long time after he died, I couldn't face playing because he was not there. I gave his clubs to a friend who plays. I knew he would use them.

I donated his clothes to Goodwill. Pictures of him are in almost every room in the house. I used to wonder if keeping his pictures everywhere was equivalent to holding my hand over the flame to see how much pain I could stand. But I can't bring myself to get rid of them, so I wish him goodnight and say hello in the morning and offer him a toast to absent love when I pour myself a glass of wine with dinner. And he smiles down at me from the top of the rolltop desk in his study while I write.

I have also made several relatively minor changes to my environment at home, mostly changes to the house. Some were changes we had made plans for before he died but had not yet gotten to (such as having the house reroofed when a leak got too serious to ignore). Some were unplanned alterations that I wasn't even really aware I wanted to make. For example, about six months after he died, I was talking on the phone and noticed a loose seam in the wallpaper in the front hall and started picking at it. Oops. This turned into a job that started out easy and quickly got much harder. I had no clear idea of what I was putting into motion when I tugged at a loose piece that afternoon, and it peeled right up, easy-peasy. It turned into a two-week project. Most of the changes were jobs that had been deferred when he was sick and my focus was on him, not on the state of the walls, windows, and kitchen cabinetry.

These surface-level changes have been needed and enjoyable despite the occasional difficulty, but they are the more "public-facing" changes I've made. The smaller but in many ways much more significant internal changes may not have been noticed by the people around me, but I have seen them, and, more to the point, I feel them.

I am rediscovering some of the pleasures of living alone, which might sound odd to some people but not to me. For the first half of our thirty-six years together, Chris and I had a long-distance relationship. His job or my job never seemed to be in the same place and were often not even in the same state, so each of us lived alone much of the time. In the second half of our lives together, having someone else around all the time took a while to become familiar. I had forgotten the freedom that comes from solitude. If I want to read until 3:00 a.m., there's no one to be annoyed by the light staying on. And if I want to have green peas and a Klondike bar for dinner, I can, without having to consider making something not so strange for someone else. I don't recall having difficulty living alone during the several years we lived apart. It can get lonely, but then nothing is perfect, and you can be lonely even with someone else there.

That first summer after Chris died was absolutely bleak. Classes were over; I wasn't teaching summer school; and I had tremendous difficulty focusing on writing and was scared that I wouldn't be able to finish the book I was working on when he died. I felt as though I were in stasis, just hovering over my life, not really a part of it or of the lives of the other people around me. I was so grateful when the fall semester started up (finally!) and I had something to focus on.

When he died, I turned to my mother for advice and solace. Mom had buried two husbands, and if anyone knew how to "do" this grieving thing, I figured she did. The advice she gave me was simple. She said, "You need to find something to look forward to." At the time, I told her that I was "looking forward to having something to look forward to" and got the laugh I had hoped my deflection would generate. But the second summer was a bit better, and they've been getting progressively easier over time. I don't feel so disconnected anymore, and I find that I do now have plans that I look forward to.

The adage I quote at the beginning of the chapter, attributed to several different authors in one form or another, has proven to be true. "Don't look back. You're not going in that direction." What is behind us is what got us here, but we have already learned those lessons. The direction that we choose now moves us forward to new lessons and whatever it is that comes next.

Perhaps best of all, I don't feel guilty anymore when I laugh. I am able to allow myself to feel happy, amused, silly, even joyful without that twinge of "Oh, how could you do that?!" sounding in my head. That voice is difficult to ignore, but gradually it gets quieter.

COUNTERFACTUALS, RUMINATION, AND WORRY

Big changes alter our world, and the bigger the change, the more difficult the struggle to adapt to it can be. When we look back on what happened, we often engage in what psychologists call *counterfactual thinking*. Counterfactual thinking happens when we generate alternatives to reality, often after a negative event, such as losing a loved one. When we imagine an alternative to reality that is better than what actually happened, we're engaging in *upward counterfactual thinking*. The opposite kind, called *downward counterfactual thinking*, involves imagining an alternative reality that is worse than what actually happened. Upward counterfactuals are probably more common after the death of someone we love because there isn't really an outcome that could possibly be worse than the loss that comes with death.[1]

When I think back to the sound I heard but didn't investigate on the night Chris died, I often engage in upward counterfactual thinking. Imagining that I had gone to see what the sound was all about, that had I called 911 sooner, that I was there with him so he didn't die alone shows me alternatives to reality that are better than what actually happened. Now, several years later, I still imagine that if I had gotten to him at the onset of what was a massive heart attack, I might have called for help sooner, but I also know it's unlikely that I would have been able to get help in time to save him. I will always wonder. I might also have doomed him to an outcome I know he would have hated, lingering in the hospital, dependent on machines, dragging out the end. A different outcome, to be sure, but not necessarily a better one.

Just before Chris came home from his last visit to the Cardiac Care Unit at the hospital (several months before he died), one of his cardiologists came to his room to talk with us about what might happen next. One of the options was some kind of ventricular-assist device that would help his heart pump. He showed us a picture of a man wearing the device, strapped to his chest, big smile on his face, happy as could be with the machine that was keeping him going while waiting for a heart transplant. Chris went pale (I'm sure I did as well). I was trying to process the words *heart transplant*, thinking, "Wait a minute. What? Heart Transplant? How did we get here?"

When the doctor left, Chris and I talked for a long time about how much he didn't want this possible end, how it was not what he had in mind. He asked me to make sure it wouldn't happen. I was just terrified at either option—being able to make sure it didn't happen and not being able to. Looking back now, I think that in many ways this counterfactual outcome would have been worse, and I'm strangely grateful that I was never in the position of having to make that very difficult decision.

Adapting to the enormous changes that his death created means, in part, being able to forgive myself for not knowing that Thursday was his last day, for deciding the sound was unimportant and choosing instead to let him get the rest he needed. It has taken a while to get to that point.

Counterfactual thinking happens to everyone, regularly. Focusing all of our cognitive energy on what might have been is not, however, helpful. Repetitively and passively focusing on negative events in the past is called *rumination*.[2] Rumination is seen in several situations, not just bereavement. We tend to ruminate when we're sad or when we're blue or when we're concerned that we've made a big mistake or when we're experiencing actual depression. It is the repetitiveness of rumination, the difficulty in turning it off, that can lead to trouble. It can keep us mired in the past, looking backward instead of forward. Researchers have suggested that rumination may make us more vulnerable to depression, and rumination about the death of a loved one can make us more vulnerable to complicated grief (described in chapter 1), thus making adjustment to the enormous changes created by loss even more difficult.[3]

Thinking in the other direction—that is, focusing our thinking on events in the future that might have a negative outcome but of course have not yet happened—is called *worry*. The death of a loved one and the enormous changes it causes does lead to more worry, as do times of uncertainty in general. There is actually more to worry about—finances, mortgages, relationships with others, and our health, just to name a few. We often worry about what might happen next in order to try to avoid negative consequences or to try to gain some control over events to come.

However, worrying without end—that is, engaging in "prolonged and inflexible" worry—can make adjusting to the loss more difficult.[4] Think back to the dual process model of healthy coping with grief discussed in chapter 2. It proposes that we oscillate between focusing on the loss of the person we loved and focusing on secondary stressors in our lives.[5] Rumination focuses on the loss itself or has what the model calls a *loss orientation*, whereas worry focuses on the secondary stressors or has what is called a *restoration orientation*. If we cannot let go of either the rumination or the worry, then our ability to adapt to the loss may be impaired. There is evidence that obsessive worrying also makes the bereaved more susceptible to complicated grief with its attendant anxiety and depression.[6]

ADAPTATION AND THE HUMAN RESISTANCE TO CHANGE

I have read in several sources that human beings are "wired" to resist change, suggesting that resistance to change is part of our genetic and biological makeup. If you think about making changes in your life, you can probably see that resistance. Even changes that we *want* to make, such as quitting smoking or losing weight or starting an exercise program, are often fraught with difficulty. We start off full of good intentions to make the change, only to sputter and slow as we try to make that change a reality. The kinds of changes the death of a loved one creates affect every aspect of our lives and are unwanted. Very few grieving people want to be forced to make these changes, so it is no wonder that making them is so difficult.

Psychologists studying how we make changes, both large and small, often describe that process as happening in stages. One stage model with a very large name, the *transtheoretical model of change*, proposes four stages: precontemplation, contemplation, preparation, and action. In the precontemplation stage, we're considering the changes we might have to make. The researcher Kimberly Calderwood, in adapting the transtheoretical model to the grieving process, says that this stage in grieving might be analogous to the shock and denial often seen at the beginning of grief. The bereaved may have difficulty accepting that change has to happen or understanding the full impact the death of their loved one will have on their lives. In the contemplation stage, we begin to accept that change is necessary and that the process

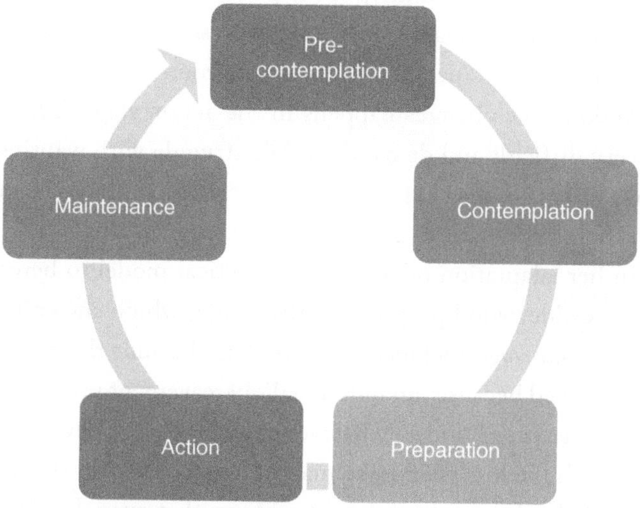

FIGURE 6.1. The transtheoretical model of change.

Source: Image from Wikimedia Commons, https://commons.wikimedia.org/wiki/File:Transtheoretical_model_of_change-_815_PDSA.png, accessed March 14, 2025.

of change may take longer than we previously thought. Calderwood describes this stage as characterized by moments of peace interrupted by overwhelming emotions as we swing back and forth between the first two stages, still not ready to make the changes we're starting to see will have to be made.[7] Figure 6.1 illustrates this model of how we make changes in our lives.

The third stage, preparation, comes with recognition of the need to make changes and that the time has come to implement them. Calderwood describes several ways in which preparation happens, which may look very familiar to you. Sorting through the belongings of the person we've lost, donating some and keeping others, collecting images of the deceased in an album or scrapbook, journaling, selling the house in which the loved one lived, and creating rituals to remember the lost person. Letting go of the physical reminders of the person is often very difficult to accept and can create anxiety and depression, but it does mark more of an acceptance of the changes to come.

Making the changes happens in the action stage. This can be very difficult and demanding. We struggle with wanting to move on, make the changes, and go forward, while at the same time being reluctant to give up the past.

In her adaptation of the transtheoretical model to bereavement, Calderwood proposes a fifth stage, which she calls the *maintenance stage*, included in figure 6.1. She adds this stage for the bereaved because "an individual's bereavement process typically never ends," and so whatever strategies we've developed to cope with the loss need to be maintained.[8] Grief and the memories of the deceased can be set aside so that everyday life can go on, but those memories are always there and can always be accessed.

The bereaved oscillate between these stages and can be in more than one stage at once, experiencing their components simultaneously. Calderwood suggests that it might be best to see the transtheoretical model in bereavement, unlike traditional stage models, as more of a continuum rather than as a series of "rooms" we move through one at a time, closing the door between rooms as we move forward. There is no closure in the sense of a locked door behind us at the end of the process. Going back through the rooms might always be a possibility.

Why Do We Resist Change?

Evolution has designed our central nervous system to be "plastic" or adaptable. When neuroscientists say that the brain is plastic, they mean that it is flexible and can change the way it works by reorganizing its structure or the connections between cells, usually as the result of an experience we've had. We demonstrate this neural plasticity whenever we learn something new. That process of rewiring the brain happens over time.

It can take quite a long time, anywhere from 18 to 254 days, to establish a habit—to learn a particular behavior and make it automatic.[9] Both establishing the habit and sticking to it require changing the connections between cells in the brain. That change is often disruptive; it happens slowly, usually involves loss (if nothing else, we lose the previously learned behavior), and can be quite painful. As a general rule, we tend to prefer what we know, what we've already learned, and what is familiar and to reject the unfamiliar that we don't know anything about.

The professor of psychology R. Nicholas Carleton has proposed that the fear of the unknown is the most basic and fundamental fear we have. It is the fear that underlies all other fears. Carleton says that fear of the unknown is caused by lack of important or key information. When we don't know what is going to happen, we feel threatened and afraid.[10] Loss of a partner throws us unwillingly and abruptly into the unknown, and part of the difficulty in adapting to the grief we feel is that we are both threatened and afraid.

C. S. Lewis opens his book on grief with the line "No one ever told me that grief felt so like fear."[11] That awful, fluttery feeling in the stomach, the restlessness, the sense that some unknown terrible something approaches, is fear. Other writers have echoed this feeling. "Maybe grief doesn't just feel like fear, maybe it is fear," writes Samantha Stein. "Grief is not just loss . . . it is also about becoming untethered. It's about losing an identity. Losing a map and compass all at once—a way to orient our life. Our love."[12]

Finding both our new self and our way in this new and unknown place takes time. The process is scary, demanding, and difficult. It is grief. Couple this reluctance to change with the demands we all face to continue dealing with everyday life, and you can see the emotional, cognitive, and physical disruption that loss creates.

Resistance to Change in the Brain

Loss of a loved one is abrupt and massively disruptive. That loss changes absolutely everything in your life. To say that this loss creates an alarm signal in the brain is an understatement. All the networks in the brain react to this alarm. The limbic circuit in the brain reacts to any change as a threat, signaling in particular the amygdala and the prefrontal cortex that something has gone seriously awry and that an important focal point in our lives around which we've learned to organize our behavior is now missing. This is an emergency, and the changes in our behavior reflect our attempts to cope.

We wander, we look for that person everywhere, and we yearn for that person, despite knowing (intellectually at least) that there is no way for the person to return. Yearning involves activation of other parts of the limbic system as well as the salience and attention networks in the brain. The salience network helps us determine what out of all the things we encounter in the world is salient or important. The attention network then focuses our attention on that important thing. In terms of grief and grieving, we might see the lost person everywhere even though we know they are not present—each "sighting" of them grabbing and holding our attention. We focus on any clue that the person we've lost might be present, might be the source of the sound at the front door, might be that flash of someone with the same hair, a similar smile. Gradually, as we adapt to the reality of the loss and the impossibility of reunion with the lost person, our attention shifts again.

Peter Freed and colleagues proposed that the *nucleus accumbens* (called the "leaning nucleus" because it looks as though it is leaning on a nearby structure) and the amygdala, both parts of the attention and salience systems, might be involved in learning our new reality and adapting to the changes in that reality (see figure 6.2).[13] The nucleus accumbens is involved when we learn something new, when we feel rewarded by the outcome of

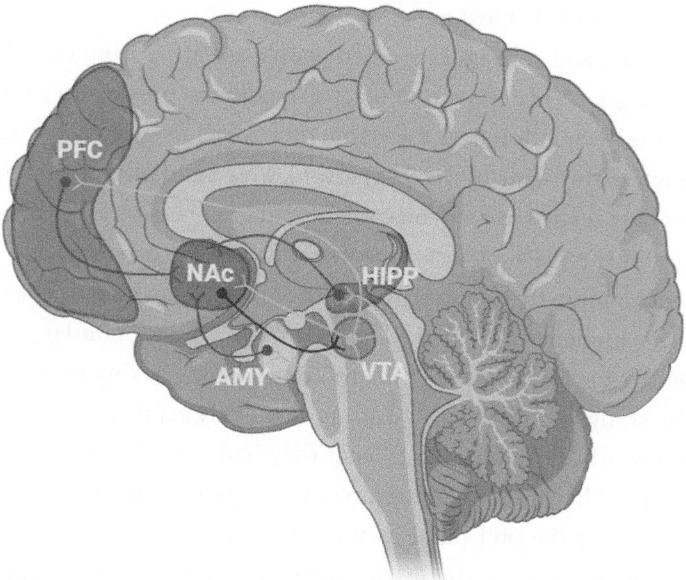

FIGURE 6.2. The nucleus accumbens (NAc), amygdala (AMY), and prefrontal cortex (PFC). The other two areas identified are the hippocampus (HIPP) and the ventral tegmental area (VTA).

Source: Image from Wikimedia Commons, https://commons.wikimedia.org/wiki/File:Mesocorticolimbic_Circuit.png, accessed March 14, 2025.

our behavior. It is also critical in memory. The amygdala plays a role in feelings of fear and sadness as well as in emotional memory. It has also been implicated in what is called *social threat*, or threats to the important social attachments we've made with other people. The death of a person is the ultimate threat to that attachment.

Studies have linked the nucleus accumbens, the amygdala, and the regions of cortex that regulate them (the prefrontal cortex and the anterior cingulate cortex, or ACC, described in chapter 2) to feelings of yearning for the return of the lost person. These regions are also tied to the sometimes intrusive memories of that lost person.

Freed and colleagues studied people suffering from complicated grief, asking them to react as quickly as possible to either neutral words (such as the name of a color) or grieving words (such as the name of the lost person). Freed and other researchers have tended to focus on complicated grief because it lasts longer (so the yearning lasts longer and is easier to examine) and is more intense than typical grief. Theoretically, the kind of yearning experienced in typical grief would be very similar, just harder to catch in research. The attention of grieving people in the study was more easily and more quickly captured by grieving words, or reminders of what they had lost, than by neutral words. The bias in their attention was correlated with increased activity in the prefrontal cortex, ACC, amygdala, and nucleus accumbens. Interestingly, the deeper and stronger the feelings of grief, the lower the strength of the connections between the parts of cortex that would normally regulate regions such as the amygdala. It was as if the normal regulation of the emotional response and emotional memories was weakened. More control over attention and decreased sadness overall was paired with greater regulation of and by these regions of the brain.

Thus, immediately following the loss, the limbic, attention, and salience circuits get activated. All three circuits affect memory, what we perceive as important and so what we pay attention to in the environment around us, and emotion. Mary-Francis O'Connor and colleagues propose that reminders of the person we've lost activate these attention, salience, and memory systems. Over time, the circuits become less activated when we encounter these reminders.[14] For people suffering from complicated, prolonged grief, though, this decrease in activation takes place more slowly or even not at all, keeping those feelings of yearning and loss front and center for a long time. However, the same circuitry is activated when the bereaved suffering from even typical grief yearn for the lost person. Our brains are searching for that central pillar in our lives.

RESILIENCE

Psychologists often refer to an aspect of human personality called *resilience*. Resilience is defined as "the ability to adapt successfully to adversity, stressful life events, significant threat or trauma."[15] Resilience is learned. We acquire resilience with repeated experiences of needing to adapt to something that happens to us, such as stress and grief. Each experience creates practice with being resilient, which in turn leads to stronger resilience the next time we have to adapt.

Resilience is probably best conceptualized as a continuum.[16] There is, however, a limit to our ability to bounce back. At some point, our ability to adapt and be resilient will crumble. Make that stressor severe enough, and we all have difficulty responding with resilience. There are few experiences more stressful than the loss of a life partner. Repeated bereavement, or cumulative grief, may be another of those severe stressors that threaten our resilience. All the changes that we experience, both good and bad, shape us.

The key to successful resilient response seems to be that the trauma is manageable. Being able to manage stress strengthens our resilience by modifying the prefrontal cortex.

Remember that the prefrontal cortex is part of the executive control system, responsible for attention, planning, problem-solving, abstract thinking, and flexibility (the system's organizational functions) as well as for self-control, regulation of our emotional responses, decision making, and initiating an action (the system's regulation functions).[17] It appears that the prefrontal cortex is part of the system in the brain that detects whether we have control in a given situation or not. If a stressful situation is seen as controllable (manageable), then that cortex's activity modifies activity in other regions of the brain that are in charge of our stress response. In a controllable situation, the prefrontal cortex can turn down the stress response we have to an emergency, and that turned-down response seems to last for a very long time (weeks or months).[18]

Seeing yourself as able to deal with stress or trauma goes a long way to actually making you more resilient. Having been successful at managing stress in the past can help with new stressors that come flying at you.

There is also a genetic component to resilience. Resilience has been shown to be what geneticists call *moderately heritable*. Heritability is a measure of how much our genotype (the genes that we've inherited from our parents) influences our traits. A statistic called h^2 can be calculated to measure heritability, ranging from 0 (meaning there is no genetic contribution to the trait and that all variability in the trait is due to the environment and learning) to 1.00 (meaning that the trait is determined wholly by genetics). A moderately heritable trait such as resilience is equally influenced by our genes, our environment, and our experiences.

Genetic studies have focused on a protective variant (allele) of genes that modulates a chemical in our brains called *norepinephrine* and how it responds to stress.[19] Other studies have found alleles that influence the activity of the amygdala and the hippocampus (involved in the creations of new memories) in response to a threat. Genes interact with the environment to shape the function of the neural system that responds to stress and produces resilience—or sometimes doesn't.

GRIT AND GRACE

Another aspect of human personality that factors into our response to loss is referred to as *grit.* Grit is defined in psychology as a personality trait characterized by persistence, perseverance, and the desire to achieve a goal.[20] Grit persists even when the going gets tough. We all probably have grit to varying degrees, and because we all want different things and have different goals, that grit can be expressed uniquely in different individuals. A part of adapting to loss is persistence. When we move forward even though we don't want to, that's grit.

Another idea often linked with discussions of grit when you read about bereavement is *grace*—of finding grace as you grieve. Grace can be defined in a number of ways, many of them much more theoretical and theological than I want to be here. I think of grace in this context as forgiveness—forgiving yourself, accepting yourself and the pain and heartache that you feel with gentleness and caring. When you have to keep on keeping on and you don't really want to, when you'd rather just curl up in a ball and sob, grace can help soothe the way.

Kristen Meekhof writes about her experiences with grief and the loss of her husband. I haven't found a better version of what she wrote about grit and grace in grief, so I'll just let her say it. Meekhof defines grace as "a feeling that envelopes you with a maternal gentleness and warmth that makes you feel even if just for a moment that you will be okay." She goes on to say that as she grieved the loss of her husband,

> I found myself seeking moments of grace. I discovered grace in the most unlikely places. I felt it in a yoga class. I found it in an email from a close friend. I saw it in the gratitude journal my late husband wrote in even during the medical crisis. While there is no stunning announcement I'm making, I'm here to tell you that both grit and grace as painstaking as they sound can help you in your journey with grief. They don't provide an elaborate affair as you won't suddenly feel complete or healed, but what I've learned along the way, is that healing occurs in small micro-moments. When you become open and you can allow grit and grace to enter, even for a moment, there comes a break in the darkness[21]

I came across this description of grit and grace in conjunction with another quote that I also liked very much. Tyler Kleeberger, writing about how slowly and painfully change happens, quotes the writer Anne Lamott, who says, "Grace is not a run, and it usually isn't even a walk. Often, grace is simply a slow scooch across the floor."[22] That image made me smile. Scooching. Changes like the ones produced by grief are painful, often discouraging, and frustratingly slow. Yet we still move forward. So, go ahead and scooch until you can walk and look for those moments of both grit and grace along the way.

JOURNALING: A WAY OF FINDING GRACE?

Writing about the loss of Chris has helped me—not only the writing but also the possibility of sharing with others what I've learned. If you're struggling with grief, I highly recommend keeping a journal. A great deal of research has been done on the psychological and physical health benefits of journaling. A study done in 1986 found that asking college students to write down their feelings about a traumatic event (not necessarily the death of a loved one) or to write just about the facts of the event, keeping feelings out of it, helped reduce blood pressure and improved long-term health.[23] A review of the studies on the benefits of writing down our emotions found that these studies showed journaling has benefits to physiological well-being and functioning as well as to overall or general functioning day-to-day.[24] Studies that examine the benefits of journaling specifically for the bereaved have shown these same benefits over the long term for this group as well.[25] Write it down. It helps.

7

FINDING MEANING

The only constant in life is change.

—Heraclitus

In the course of writing this book, I've talked with a number of people who are grieving the loss of someone they loved. One question I often get, both from the bereaved and from people who love them, is about how long recovery will take. When will I be normal again? I asked the same thing when Chris died. The bereaved want to know how long they will have to deal with the agony of grief, and the people who love them want to know that the grieving person they love will be happy again. I struggle with how to answer this question because I don't want to throw cold water on their hopes—hope and its cousin, optimism, are important—and because there is no simple answer. Yes, the pain of grief lessens, and moving forward becomes a possibility. But I also have found through my own experience that grief, the sadness over the loss, probably never ends. The hole the absent person leaves behind will always be a feature of life going forward. However, I also think that any satisfactory answer might just depend on how you define recovery.

HEALING AND RECOVERY

Let's address what we mean when we say "healing" and "recovery." The term *healing* usually refers to becoming sound or healthy again. Most people use the word *recovery* to mean essentially the same thing, returning to a previous normal state of health, mind, or strength. Most people who experience a physical loss recover and heal (although, of course, there are situations in which they do not). Their bodies deal with the stress of loss and regain equilibrium. Their minds cope with the disorientation and confusion that loss has produced and regain an even keel. However, many people who study grief and grieving will tell you that loss like this changes who we are and how we see and understand ourselves. In fact, several authors have pointed out that words such as *recovery*, *healing*, and even *resilience* may be inappropriate to describe the bereaved and have offered several alternative descriptors, proposing *adaptation*, *reintegration*, *management*, *adjust*, and *cope*, among others, instead.[1]

The debate about which word is appropriate stems from a more fundamental disagreement about what the point of grieving might be among clinicians and counselors who seek to help the bereaved with this difficult process. Older theories suggested that the purpose of grief was to disconnect ourselves from the lost person, and the hoped-for outcome of grieving was a return to our previous state of being—in other words, we experience recovery as most people understand it.[2] The newer theories have come around to seeing that loss fundamentally changes the way the bereaved define themselves and acknowledge that there is no going back to the person we were before the loss; that person is gone.

To me, that sudden abrupt loss of such a major focus in your life is like having an arm or a leg torn off and being flung out into the future, minus that essential part of yourself. You are in pain, disoriented, lost, confused, and you discover yourself in a completely new place, on a completely different path in life. You have to figure out who you are now, without that limb, without that focus, in order to determine who this new you is going to be. When you get that figured out, you have to relearn how to regain your balance without that missing limb and move forward again in this new place. That is not "returning to a previous state" at all. This new and different you is the person who has to move on. That's what researchers and writers mean when they say that grief never goes away. Grief makes you a different person. Your job now is to learn to live as that new you.

Grief profoundly alters the person experiencing it, to the point where there is no going back to being the same person before the loss. You can, however, be healthy, happy, and "healed," moving forward as a new and different person. Remembering the lost person, wishing they were still here, missing them, yearning for them do not mean you have not healed, adapted, or moved on. It means you loved that person. It means that you have been changed by their loss. It means you are human. Thomas Attig addresses the often unwelcome durability of grief by saying, "We eventually dissipate or overcome some of our pain. We learn to carry some other of our pain [*sic*]. But, more important, we move from *being* our pain—being wholly absorbed in and preoccupied with it—to *having* our pain—to carrying residual sadness and heartache in our hearts."[3]

Part of the process of moving forward involves getting used to all the changes that have been forced upon you, and part of it is figuring out what being this new person means. You must find meaning in what has happened and in what you've become.

THE STORIES WE TELL OURSELVES

When we lose an important bond, we are driven to make sense of that loss. Reminders of that loss trigger activity in regions of the brain that are involved with memory, what we know about the world and the people in it and how they interact and how we interact with them. In all memories, we knit together stories of what has happened. Narrative making is the process humans use to make sense of the events around them, both profound events, such as loss, and ordinary everyday events, such as making a date to meet up with old friends.[4] When we've lost someone, grieving involves reconstructing our narratives that involve that person. The stories we create have a beginning, a middle, and an end, and they are how we organize and understand the events that shape our lives.

Along with everything else the grieving struggle with, there is often this sense of the loss of meaning in life. Regaining that sense of meaning can take a long time, but the process can be a positive one. The new self you are discovering isn't automatically worse than the old you. In many ways, it might be a better self than it was before the loss.

Creating the story and rewriting that story when the unexpected or traumatic happens require us to make connections among events, people, relationships, and the words we use to describe them. Making those connections creates meaning.

A number of models describe how humans figure out what life means and how they go about changing that meaning when life throws a curve ball at their heads. I won't go into depth on what each of these models suggests about how we search for meaning after loss. Instead, I'll focus on what I and others think these models share.

The bereaved are essentially faced with a major challenge to what psychologists call their *assumptive worldview*. We all have and develop these assumptions through experience with life and what we know about how the world works and who we are in that world. This assumptive worldview is a story or a series of stories that we've created over time about ourselves and the life that we live.

Loss creates a major discrepancy between what we assumed life was going to be like before loss and the reality of life after loss. The clinician Robert Neimeyer says that "meaning reconstruction in response to a loss is the central process of grieving."[5] Losing someone we love requires that we rework the stories and memories we've created over the years about how every aspect of the (our) world operates. This reconstruction of our narratives happens, says Neimeyer, "at every level from the simple habit structures of our daily lives, through our identities in a social world, to our personal and collective cosmologies, whether secular or spiritual."[6]

These models say that when something unexpected and outside of our control happens, we are faced with a situation where how we understood the world before the loss no longer matches what we're experiencing. Something has to change. Either we figure out how to assimilate our experience of loss into an unchanged worldview, or we change our worldview. Something has to give.

ASSIMILATION AND ACCOMMODATION

If you've ever taken a psychology class, you may remember a developmental psychologist named Jean Piaget and his description of how our cognitive abilities change with age. In his model, Piaget used the idea of a *schema*—a way of organizing information—to refer to both what we know and how we acquire that knowledge. There are different kinds of schema—self schemas, or our knowledge of ourselves, who we are, and who we want to be; social schemas, or knowledge of how people behave in social situations; even event schemas, or understanding how we should behave in a particular group situation.

We go into a new situation with a schema for how that situation might work. Sometimes that schema is elemental and simple; sometimes it is very well developed and elaborate. If we encounter new information, we try to make it fit with what we already know, with our existing schema. That new information might require some tweaking, some small changes in it, to make it fit what we already know. That process of fitting the new in with old is called *assimilation*. We assimilate that new experience into an existing understanding of how the world works. If the new information does not fit with an old schema and cannot be assimilated, that old schema is modified. Changing an existing schema so that new information can fit into it is called *accommodation*. It's often easier to assimilate than to accommodate, even though assimilating the new might require changing it a bit.

We need these schemas to be coherent and to explain the world adequately. We use these schemas to understand the world, to predict what will happen, to feel in control of what is going on around us. The older we get, the harder it is to change our schemas. It can often take a massive amount of new information or severe trauma, such as the loss of a loved one, to get us to change a schema. The enormous changes this loss creates make us reconsider our existing understanding of the world.

The pain of grief can force us to look for meaning in the loss and, as a result, to assimilate and accommodate the disruption in our assumptive worlds. Rachel Coleman and Robert Neimeyer say that "in the aftermath of personally devastating loss, survivors strive to adjust or reconstruct their assumptive worlds. . . . [B]ereavement prompts efforts to find meaning in the troubling transition, with new meaning being retained and integrated," assimilated and accommodated, "to the extent that they [such efforts] reduce distress; otherwise attempts at reconstruction are likely to continue."[7]

THREE WAYS TO FIND NEW MEANING

Most of the models of how we find new meaning after loss suggest that there are three ways the bereaved try to construct this changed worldview and rewrite their narratives: (1) by making sense of what happened, (2) by understanding the change in their identity, and (3) by finding benefit in what happened.[8]

These three methods for rewriting our narratives involve both assimilation and accommodation. We assimilate our understanding of the death of the person we love into our perception of who we now are and where we are in our lives, making whatever changes are needed to that existing sense of self to fit in the new information. We come to accept the fact that our important person is gone, to accommodate our assumptive worldview to this monumental change in how we understand the world, and to adjust our sense of identity and our relationships to others [9]

The third method, finding a benefit in what has happened, often makes people recoil a bit—I know it did for me, at least initially. How can the loss of someone we love possibly be beneficial? It can be hard even to think about because it can feel like a betrayal of that lost person. But, yes, there are benefits to be had, and they are more profound than the tiny benefits we might first think of, such as smaller grocery bills and more room on the bathroom shelves.

In a study done in 1998, a large group of bereaved individuals were asked specifically and directly about the positive aspects of grief and grieving and what they'd learned about themselves through grieving the loss of an important person in their lives. The benefits were often the change in self-knowledge that came about because of the loss. People reported that their relationships with family and friends had been strengthened by grief, that the loss had given them a better appreciation for life and helped them get their priorities in order, that they were better able to rely on themselves, that their fear of death had diminished, and that their religious beliefs helped them see the lost person as "in a better place."

Many of the people interviewed said that the burden of caring for their beloved person had been lifted from them and that they had learned patience and compassion for others as they grieved. They also reported that they had learned to stop and smell the roses and not to let go by any opportunity to tell friends and family that they are loved.[10]

EXPERIENCE CAN BE A HARD TEACHER

The question of just how positive or negative experiences change us has been the subject of research in psychology for a long time. One of the main questions researchers have been asking is about whether pain is required for change to happen. Must the event be a horribly painful negative event such as loss to engender change, or can positive events, events that don't hurt, also change the way we see ourselves?

In 2018, three researchers from Germany collected and examined 122 studies that had examined growth after a negative event (*post-traumatic growth*) and after a positive event (*postecstatic growth*). They found that social relationships were reported as stronger after negative events compared to relationships after positive events, which suggested that improving social relationships might require suffering. However, self-esteem improved after both positive and negative events, and the feeling of mastery over the environment improved more after a positive event than it did after a negative one. Feelings of personal strength were improved after a negative event (there were no studies that looked at the effect of a positive event on this aspect of our self-assessment). Other aspects of our sense of self—such as feelings of spirituality and the sense that we have a purpose in life—did not appear to change after either negative or positive events.[11]

HOPE, OPTIMISM, AND RECOVERY

Take a look at the history of science, in particular the history of psychology, and you'll notice a pattern. There is a tendency for psychological theory and research to focus on what can go wrong in our mental lives as well as in our physical lives. A former student of mine described introductory psychology courses as "trainwrecks 101" because of the dominance of disease, disorder, and destruction in the material. The description sort of makes sense because in psychology there is a natural interest in what can go wrong. To fix what has gone wrong, you need to understand what it means for mind and body to fly off the rails.

That tendency to focus on maladaptive and disordered thinking and behaving made the appearance of what is known as "positive psychology" surprising. Positive psychology was "born" in the late 1990s and early 2000s, but elements of it can be traced back to the post–World War II 1950s and even, some argue, all the way back to the beginnings of psychology in the United States and William James in 1906.[12] Positive psychology emphasizes human strengths rather than human weakness, focusing on what can go right in our lives rather than on what is going wrong.

The study of hope and optimism and how they affect our bodies and our minds is a good example of the research focus of positive psychology. Optimism and hope are related to one another, but they are not the same thing. Optimism is a general expectation that things will go the way we want them to in the future, that they will work out in the end. Optimists focus on general expectations, not specifically on how those expectations can be achieved. Hope is where we see objectives or goals for the future, coupled with plans to achieve those goals and what positive psychologists studying hope call "agency" or motivation directed toward achieving those goals.[13]

With regard to grief and grieving, both hope and optimism have been linked to better recovery (so, despite my earlier vacillating on the use of this term, I'm going to adopt it here as shorthand). In general, bereaved or not, people who are hopeful and optimistic tend to report less distress in their daily lives than do pessimists and the unhopeful. A study of cancer patients found that the higher these patients scored on measures of hope and optimism, the less depression and anxiety they tended to experience.[14] Other studies have found that hope predicts greater satisfaction with life, greater happiness, and even better performance in school.[15] You can even teach people to become more hopeful. Researchers found that although hope is determined in part by genetics, it is also influenced by the experiences and environments that surround people in everyday life. Helping people achieve their goals, helping them succeed in facing obstacles and stressors, providing social support when it's needed, and even just giving them the opportunity to hang out with hopeful people can boost hopefulness and reduce anxiety.[16]

Perhaps most importantly, the absence of hope and optimism makes the bereaved more susceptible to mental distress, such as anxiety and depression, and to complicated grief. Sufferers of complicated grief are stuck in their grief. Their focus is on the lost relationship, the memories they have of their beloved, the past rather than the future. They yearn for their lost person so intensely that this preoccupation interferes with daily life, creating the feeling that they've lost their purpose in life. Their thinking is focused on rumination and thoughts that begin with "if only" (remember that ruminating is obsessive preoccupation with the past). They also tend to worry about a catastrophic future without that person and do not think about accommodating or assimilating the reality of that loss into their present lives. Complicated grief is relatively rare (estimates are that about 7 to 10 percent of the bereaved develop it), but people suffering from this intense and long-lasting form of grief can be helped by counseling.[17]

If you're having trouble looking forward with any kind of hope or optimism, talk to someone. I am not going to offer explicit advice, counseling, or therapy here because I'm not qualified to do so—my work in psychology and neuroscience is not with clients or patients. I work in the lab. But if you need to talk with someone, plenty of trained people are available to help. Find someone you can talk to. Talk to them. There is no shame in asking for help if you need it.

HOPE AND THE BRAIN

Neuroscientists have discovered in the lab that hope can help shield the brain from the effects of anxiety. A study in 2017 examined brain function using a special kind of brain scan called *resting-state fMRI*.[18] This kind of brain scan measures spontaneous activity in the brain—brain waves that happen when a particular region of the brain is just resting, not working on a particular task or objective. This technique allows researchers to see what areas of the brain are most active in people who can be described as having hope as a stable personality characteristic in comparison to the spontaneous activity patterns in those who are not "dispositionally" hopeful. It might provide researchers with a marker for the personality characteristic of hopefulness in the brain.

Previous research had noted that a part of the frontal lobe of the brain called the *orbitofrontal cortex* tends to be hyperactive in anxious individuals. The orbitofrontal cortex is part of a system that processes reward, motivation, problem-solving, and goal-directed behavior. The link between activity in this region and anxiety suggested that perhaps anxiety is the result of this system being in hyperdrive—too active, too much, all the time.[19]

In the study done in 2017, people who scored high on personality tests measuring hopefulness tended to have lower spontaneous activity in the middle part of the orbitofrontal cortex than did the less hopeful. Individuals who were highly anxious tended to have much higher levels of spontaneous activity in this brain region. Researchers speculated that higher levels of spontaneous activity in the middle orbitofrontal cortex of people who are anxious might reflect a brain working harder to suppress other areas of the brain that are creating the anxiety. In hopeful people participating in the study, this increased effort wasn't needed, so lower levels of spontaneous activity were seen in the frontal lobe.[20] Perhaps hope was protecting the brain from anxiety.

MOVING FORWARD

I see some of these changes in myself over the past six years. Before losing Chris, I saw myself as fairly optimistic and happy, able to see the positive side of most things that happened. I had a tendency to think that things would work out the way I wanted them to because, frankly, they always had up to that point. It might have taken hard work and time, but I usually achieved the goals I had set for myself.

I think of myself now, after the loss, as sadder and less optimistic but not exactly pessimistic. The optimism that had sustained me when bad things happened or life didn't turn out the way I wanted it to is now more tempered, more realistic, less buoyant. Now I know that some things that I really want are just not going to be mine. I know that no matter how hard I work, I can't force life to be other than what it is and, fundamentally, that life ends for each of us. This was something that I knew intellectually but wasn't really cognizant of emotionally, if you know what I mean. When Chris was diagnosed with cancer and heart disease, we both knew that this combination was eventually going to take him out. After he died, I saw the inevitability of death in a new way. We come with a "use by" date, only we never know exactly what that date is.

Chris's death has made me "stop and smell the roses" more, although I know that's a cliché, and to think about the moment right in front of me rather than years down the road. My relationships with other people, family and friends alike, have changed as well. I've always had a short temper, but I've found that my fuse is longer now. I'm more likely to ask myself if this thing that happened is worth a huge fuss and to take a step back and see what might happen instead of jumping in with both feet and trying to steer events as I think they should go. I also hope I've developed some patience—something I have never had in any abundance.

At the same time, I'm much less willing to put up with selfishness, the blaming of others for one's own faults, and general nonsense of the same sort. I'm much more willing than I used to be simply to end the conversation if I see this type of attitude creeping in. I had the unusual experience of walking out of a meeting that had devolved into the "blame game"—something I never would have done before Chris's death. Life is too short for this stuff.

The last chapter in this book is devoted to the stories of other people I know who have experienced loss. I was honored and gratified that so many people, both folks I met along the way of writing this book and friends I've had for decades, were willing to share their grief with me and with the readers of this book. Writing has always helped me put things in perspective, and this writing effort was no exception. I hope that sharing their stories has helped my friends in their grief. I hope they will offer hope to you as well. I will also share what I have learned and what has been comforting for me in the process of doing the research. I hope it offers some comfort for you.

Remember: it gets better, it gets easier, but talk to someone if it doesn't. You are moving forward, not backward, and there is meaning in this loss. Peace.

8

STORIES OF GRIEF AND GRIEVING

When I began the process of writing this book, I did what I usually do when I start a new writing project. I read. And then I read some more. When I finally put pen to paper, I didn't focus at first on what my questions were. I focused instead on what the researchers had to say about the experiences I was having and on trying to put their research into some kind of frame, some kind of order. I didn't really stop to think about what I was looking for until I finished roughing out what the research had told me. When I went back to look at what I had written, I could see that both my reading and writing were being driven by the questions and fears and anxieties I had about grieving the loss of my partner. I wrote this book to share what I'd learned and the answers to my questions.

I don't expect everyone's experiences with grief and loss to be exactly like mine, but I do hope that what I've learned can be of help to someone else. Along the way, I discovered that many of the questions I had were the same ones commonly asked by other grieving people—questions such as how long grief will last and, because grief brings with it feelings of anxiety, confusion, fear, and sadness, questions about what is typical in grieving. In addition, many people notice changes in themselves after losing someone important to them and want to know if the changes they see are "normal" or typical of grief and grieving.

I wanted to know what the purpose of grief might be. Does whom you lose matter? I wanted to know how long it would last, how many weeks, months, years would have to go by before I would feel anywhere close to better. How would I ever be able to look forward, to focus on the future instead of the past? Was what I was going through "normal"? Did everyone feel this way? Why am I angry? Am I angry with him for leaving me? That's not right. He's the one who died. Does this anger make me a bad person? How can this painful process possibly be beneficial? And because experience always teaches us something, what has the experience of grief taught me?

I can also say that writing about and talking about my own story and friends' stories of grief and loss were incredibly helpful in sorting out the jumble of emotions, pain, and upheaval. In the process of writing, I've come to appreciate two basic truths about grief. First, grief is both universal and individual, despite the contradiction that statement implies. Grief carries both the same awful pain and confusion of emotions for everyone who suffers from it, while at the same time it is experienced in a way that is unique to each of us. My grief probably does not look exactly like someone else's. I'm not even convinced that the specific relationship the bereaved had with the person they've lost matters all that much, either. I've never seen much value in comparing pain, in claiming that my pain is greater than, sharper than, deeper than someone else's. The pain that loss creates is a universal truth to grief. Loss and the changes we are forced to face because of it hurt a great deal.

Second, although I've come to see grief as lasting forever, it does not stay the same over time (thank goodness for that). Grief changes over time, becomes softer, comes to occupy less of every waking moment. It still comes in waves, but there is more time between peaks, and the peaks themselves are lower.

Most of what I've written about so far has been my own experience with grief and grieving. In this chapter, I want to explain what I've learned about grief and grieving in the process of writing this book. I also want to illuminate the two truths that I discovered by asking other people who have also lost loved ones to elaborate on their grief. The set of people I asked to contribute are friends or friends of friends. If you're looking for scientific objectivity here, I'm going to disappoint you. I didn't ask a random selection of the bereaved to contribute because I was reluctant to approach a stranger with intrusive questions about a very painful and private experience. It was hard enough to ask friends, even though they all graciously agreed without hesitation to tell me about their loss. I know that talking with them about their loss helped me, and I hope that sharing my talks with them will help you as well.

The set of questions I sent each person was based loosely on the questions I had about grief and grieving and conversations with friends and family. What I present here is a sort of distillation of their answers. I was fascinated to learn that I was by no means the only one with these same questions.

You will also quickly notice that almost all the people who agreed to tell me their stories of loss and grief were women who had lost their life partners. One friend agreed to talk about the loss of her son. Loss of a child is the only kind of loss to rank higher than loss of a partner in the level of pain it causes, according to people who keep track of this kind of thing.[1] Another friend touches on the loss of a sibling during her childhood, comparing it with the loss of her beloved husband as an adult.

I focused on people who have lost a partner for two reasons. First, I know so many people who have lost a partner, so this group was depressingly easy to find. And second, loss of a life partner, regardless of the gender of that partner, is one of the most stressful life events we face, which made this group of bereaved individuals good spokespeople for grief and the process of grieving. The loss of the person we've chosen as a partner in life changes everything about our lives because they featured and continue to feature so prominently in our daily lives (both past and present) and in our plans for the future. Loss of a parent, aunt, or uncle is more "expected" in some horrible definition of that term, which diminishes the intensity of the pain somewhat. In addition, in most cases when our parents die, most of us no longer live with them and have not shared their daily lives for quite a long time. The loss of a person whose everyday life was so tightly bound to our own makes the pain of their loss sharper and deeper in comparison.[2]

SOCIETY, CULTURE, FAMILY, AND GRIEF

A great deal of research exists about the effects on grief of age, gender, and the specifics of the relationship between the bereaved and the lost person. What such studies have found, essentially, is that the specifics of the loss—who was lost, the age at which we experience the loss, and so on—don't change the experience of loss very much. We still experience pain, emotional upheaval, physical distress, and general confusion with the loss of any important relationship. What varies is how long these aspects of grief last and their intensity when they strike. The social rules that we all more or less follow have an effect on exactly how we express grief, but they don't seem to affect what we feel.[3]

I think this debate about the specifics of the relationship we had with the lost person and the effects these specifics have on the grief we experience comes back to the distinction I made in chapter 1 about grief versus grieving. Across the world with all its cultures, subcultures, and societies, we all make bonds with important other people. Those bonds are essential to our survival, which is why we form them. When those people we have bonded to die, we feel grief. The exact form of that grief is unique to each of us.

The way we express that grief, the manner in which we grieve, depends on many factors. Our culture and the society we live in dictate the expectations we have about how grieving should happen, who gets to do it, how it can be shown to others, and all the rituals we engage in during the grieving process. Our gender, our age, our families, and our status in society influence how we grieve. The process of adapting to loss will not look the same in every grief-stricken individual, but the fact that the loss of the bond of love generates the need to adapt—that is universal, I think.

The next section covers the questions I had at the onset of my grief as well as both my own answers as well as the answers provided to me. I've summarized the answers in the interests of space and efficiency. If you would like to read the complete answers each person provided, they can be found in an appendix.

GRIEF STORIES

1. Whom did you lose?

If you've picked up this book and gotten this far, then you know that I lost my husband and partner of thirty-six years, Christopher Robinson McRae, to cancer and heart disease. At the time of this writing, I am six years "out" from losing him. In the years since he died, I also lost my childhood best friend and my mother. My relationship to each was different, and each loss was different as a result. For me, the difference in the grief I felt at each loss was in its intensity rather than in exactly what that loss felt like. The loss of Christopher was the most intense grief I have ever felt. There was pain, emotional confusion, and physical distress associated with each lost relationship. But his loss was different: it was a difference in quantity (intensity) as opposed to quality (exactly what I felt).

The other women I talked with had a very similar response to this question. All of them had lost a life partner, and for many of them that loss was only the most recent one. Rosemary describes the loss of a sibling: "I had lost my brother when I was a child (he was 18 and I was 10 when he was killed in a car accident). I remember the pain of that. It changed my whole life and that of my parents. But it didn't compare to losing a partner."

Another friend (LC74) describes dealing with the loss of her son, which precipitated the dissolution of her marriage to her son's father, another form of loss generating another form of grieving. The loss of a child or a life partner is life altering in intensity.

In addition, all the deaths these women describe were unexpected, even when the actual death of the person followed years of treatment for the disease or disorder that was eventually responsible. This, I think, is a feature that all losses share. In my opinion, you can be intellectually prepared for loss, but the emotional reality of it is quite different.

2. How long ago did the loss occur?

This question was a roundabout way of asking how long the pain and sorrow of grief lasted. I think this is one of the most frequently asked questions about grief for a very good reason.[4] It is awful to experience, terrible to have to endure, and terrifying to think about it possibly lasting forever. Chris died six years ago (January 11, 2018). I was sixty-two years old, planning for my upcoming retirement, and looking forward to selling the house we were then living in to move to Savannah to allow for better pestering of the kids and grandkids. That did not happen as planned. For a while, I was afraid I would never again have what I wanted, never feel normal again. I can say that after six years I feel better, but with a new and different normal.

Time, space, and emotional distance from loss help. S.R.H lost her husband of fifty-four years only six months before answering my questions about that loss. She wrote her answers without having yet been able to achieve much distance from that loss, and the immediacy of her grief is painfully apparent in her answers. She wrote that she felt his loss profoundly: "I feel diminished. I am a roller coaster of emotions. I, who like to think of myself as even-keeled, am now quick to anger. If I am with a group of people, I am out of synch with their conversation. If someone mentions his name, I think they have no right to say it so casually, but if they talk about what's going on in their lives, I wonder how they can be so callous. I am not myself. I don't know who I am."

Compare that answer to the answers from women who have had more distance. Liz (seven years out), who works with a social support organization dedicated to helping the bereaved, says that she advises the newly bereaved that the second year can be worse than the first. She says, "Reality sets in after the first year of 'busyness' with estate settlement, canceling accounts, perhaps moving. . . . I say it 'doesn't get better, it gets different.'" EC (ten years out) talks about the loss of the new adventures she and her husband were anticipating now that the kids were grown and out on their own but notes that she now looks at that loss with regret, not with the overwhelming pain she experienced at the beginning of grieving. And then there's the answer from Sheila (thirty-nine years out since the loss of her husband). Although she misses her husband, she writes about the peace she feels in the happy memories she has of him and her gratitude for having known him. The grief changes; it gentles and (I hope) becomes more peaceful. It takes time for this to happen. I just don't know yet how much time.

3. Can you tell me about the circumstances of that loss? How did your loved one die?

My husband was diagnosed with prostate cancer twelve years before his death. The most recent treatment for the cancer, which had metastasized, was a new drug that suppressed testosterone levels. Since prostate cancer feeds on testosterone, suppressing testosterone levels suppressed the cancer. The drug he was taking when he died had been linked to an increased risk of congestive heart failure, a fact we did not know at the outset of treatment with this new drug. Would that knowledge have changed what we decided to do? I'm not sure it would have. Cancer or heart disease . . . talk about choosing between the devil and the deep blue sea.

The issues with his heart began with a five-way coronary artery bypass graft four years before his death. He developed congestive heart failure about a year before he died. I don't know which lethal disease took his life. I suspect it was both. We knew that he was suffering from the disease (or diseases) that would eventually end his life. He was also thirteen years older than I, so I pretty much knew that he would most likely die before me. But his death before mine, no matter how expected, was still unexpected (if that contradiction makes sense). I had no clue that particular Thursday would be his last. I think that no matter how clearly you see such a loss coming, you still don't believe it when it gets there.

Many women who volunteered answers described losing their partner abruptly and unexpectedly. The circumstances surrounding the loss ranged from a sudden death that took place in the hospital for what was supposed to be routine treatment to returning from a quick trip to the store to get ice for a party and discovering her loved one dead on the floor to that phone call from the hospital trauma center that we all dread, saying, "You need to get here now; there's been an accident." I have tried to put myself in Edna's shoes, but I don't think I can. Her husband committed suicide in front of her as they were taking the groceries from the car into the house. The agony of that memory almost defies understanding.

The variety of ways in which death enters our lives is a large part of why I think all losses, even the loss we see coming, are unexpected. There is no way to prepare for it, and there's an enormous difference between recognizing intellectually that death is inevitable and seeing (or hearing) it happen right in front of you.

4. What is your opinion on the "medicalization" of grief? Should grief be treated like a disease?

I disagree with the notion of grief as a medical disorder. I see grief as a part of life, a part that all of us will experience at some point. I dislike the medical view of grief as a disease because it suggests a uniformity to the experience of grief that I don't think exists in real life and that we might very well be able to "cure" it. I don't think that's possible. I'm not even sure if we should *try* to cure or prevent grief. However, I will also say that I don't think there is a correct answer to this question, only an answer that is correct for the individual.

My friends were mixed in their answers to this question. Some said that they agreed with me, that grief should not be categorized as a medical disorder; others said that it should be. I think the main differentiating factor in their answers might have to do with the circumstances of the loss they suffered. S.R.H., who lost her husband to cancer, agrees with the old adage that grief is the price we pay for loving someone: "No, I don't think grief is a disease or should be treated like one. I'm not even sure what that would mean. It's part of the journey we will each of us travel, if we have been fortunate enough to have loved someone."

Edna, dealing with her husband's suicide, has a very different opinion. Edna's children advised her to seek help in dealing with this particularly devastating circumstance of loss. She did so and found that the counseling and antianxiety medications she received were helpful in getting a handle on the guilt and pain she was feeling. In her case, "I wouldn't call it a disease, but I would call it a medical condition because I don't think I would have gotten better without the medical and counseling help I received."

Rosemary, who lost her husband to an unexpected heart attack, had an answer that was somewhere in the middle. She was surprised by the physicality of grief, the whole-body feeling of being attacked by a disease. She says: "My whole body reacted.

I had never had high blood pressure, yet my blood pressure was high. I felt like my whole body was shaking. I asked my doctor for antidepressants. She explained that she did not believe that you could medicate your way out of grief. And I agree with this. Although for me grief was very physical (I felt sick), I agree that it isn't a disease."

Liz also speaks to the physical nature of grief, saying that although she doesn't consider grief a disease, it "certainly can have biological symptoms, such as 'widow's fog' for the first year (like 'pregnancy brain'—you are forgetful and need to write down important things or create to-do lists). Obviously, sleeping and eating patterns can be impacted. Fear can be an issue. Anger too. The latter two caused me problems in the first and second years out, respectively."

Just as I don't see only one right answer to this question, I don't see only one way to deal with whatever changes grief causes either physically or mentally. I immediately came down with laryngitis after Chris died, but in the back of my mind, probably because at one point in my lab career I was studying the effects of long-term stress, I chalked up the laryngitis to being just that—the physical effects of enormous stress. I sort of knew that as the immediate stress eased somewhat, so would the laryngitis—and it did.

I think there are many ways to deal with grief. Some people benefit from medical or psychological help. Others do not. It comes down to what I learned in an introductory counseling class. (No, I am not a counselor—one class does not a counselor make. I preferred the laboratory to the counselor's office.) If you want help, ask for help. If the help you get works for you, stick with it. If it doesn't, try something else. There is no shame in either asking for help or going it alone.

5. Do you see grief as having a purpose? If so, what is the purpose of grief?

This question was a toughie for me and for the women who responded to my questions. In the beginning, it was very difficult to see or imagine a purpose for all this pain. However, the more I read about grief and grieving, the more I came to think of the process of grief/grieving as absolutely essential and inevitable to our life experience. Grieving is the very painful process of change. Change is hard to do at the best of times, even though we do it all the time. But when you *have* to do it and don't want to, it can be amazingly painful.

The pastor and speaker John C. Maxwell said, "Change is inevitable. Growth is optional." The problem is that this change, this grief, hurts, and we want it to stop. I've got a variation of this quote in my living room. It is allegedly from the Dalai Lama, and it reads, "Pain is inevitable. Suffering is an option." To me, what this means is that being forced to change when you don't want to is just plain painful. But lingering in that pain—staying there and not trying to move forward from it—is just suffering, and you don't need to do that. I try to remember this on bad days.

This question bothered me the most as I was reading the research. I've been involved in science all my adult life, and as a result I was very familiar with Darwin's theory of evolution, which serves as the foundation of modern psychological and biological science. We are shaped both by what happens to us over the relatively short time spans of our individual lives and by the forces that have acted upon us as a species over millennia. I knew from my personal experience that grief created illness, that it made me see myself as out of control, anxious, fearful, and alone, and I came to realize that it would be possible to die from a broken heart—a scary realization. Hardly a boon to the survival of the species or to my own. I suppose that the purpose of all this pain and anguish is to alert us to the need to change, whether we want to or not, to motivate us to move forward, even if it is with a limp and a whimper, and to adapt to what has happened.

Once again, I think that time and emotional distance from the loss matter in how my friends answered this question. S.R.H. wasn't sure she had an answer at all, and it may well have been too soon (only six months away from the death of her husband) for her to try to find one. Edna, struggling with her husband's decision to end his own life, says that while grief may be the price we pay for loving someone, "I just wish there was a faster and more pain-free way of getting through my grief." Rosemary, three years out from her loss, says that she wishes she knew the purpose of grief, but she isn't sure what it might be. Liz sees grief as getting us started on reshaping our lives after having someone ripped away from us, and EC (ten years out) says she isn't sure there is a purpose to grief but that it is a reality of life that many of us are unprepared for. Catherine (twenty-five years out) says: "I can't see how dying suddenly of a heart attack at age 41 could ever make sense. It has no purpose. And because the death had no purpose, I can't see how the paralyzing despair, anger, fear of dying, fear of everyone around me dying could have purpose other than pushing me down so far into a deep, dark hole that I didn't think I could ever make it out alive."

I'll give LC74, twenty-eight years after the tragic loss of her son, the last word on this question. I like her answer. She sees grief as a life lesson, one that should help us help others in the same spot: "No, I don't believe there is a purpose of grief; rather, one learns how to deal with grief, and it is our purpose to share with others going through similar circumstances and help them to understand what has happened in their life and offer support and care."

6. Do you think that your sense of self has changed? Are you a different person after the loss? What about you has changed?

Modern theories of grief and bereavement suggest that the pain and sorrow of grief are part of the sudden and usually unexpected nature of the loss of a loved one and our abrupt reorganization of our sense of self impelled by the loss of that person. Grief and loss were unwelcome experiences in my life, but I hope that I grew from them. I know I changed. My father died when I was twenty-six years old. He was someone I thought would be a constant in my life. Like many young people, I sort of assumed I was immortal, and that immortality extended to the people that I loved. When Dad died unexpectedly (he was only fifty-seven years old), I learned that no one is constant in a life. No matter how much you love someone, they will not be there forever. My understanding of the world as a safe and at least somewhat predictable place was profoundly altered through loss. Life was no longer particularly predictable and is not safe.

When Chris, then Patty, and then Mom died, that lesson about safety was underscored for me, but in a strange way also muted somewhat. The biggest change in my sense of self is that my sense of my own mortality has been altered. My fear of death has diminished as well. (It didn't disappear; it just diminished. I'm still not ready to go yet.) I know that the world is not safe, but I also know that hiding from life because it can be dangerous is pointless. The most painful change was in my sense of having a partner in the hot mess that is life. I no longer have someone to experience these things with, and that is sad, scary, and lonely. But I also know that I can deal with this.

The intense pain is one of the essential changes that grief creates is readily apparent in S.R.H.'s response when she says: "I am not myself. I don't know who I am." Edna says that of all the changes in her life, the ones created by the death of her husband have been the hardest to deal with. She never envisioned herself in the role of "widow" and is having to learn quickly how to be in that role. Rosemary says that her sense of herself as an individual, no longer part of a couple, is the biggest change. She quotes the writer Anne Lamott, who said that being widowed is "like having a broken leg that never heals perfectly—that still hurts when the weather gets cold, but you learn to dance with the limp."[5] "I'm still learning to dance again," says Rosemary. As one of the biggest changes in their lives, Liz, EC, Catherine, and LC74 also write about learning how to no longer be part of a team, a couple, and about abruptly needing to learn how to be alone again and do things solo. It certainly has been a significant change in mine. It requires rethinking and reimagining every aspect of your life, and that is difficult to do.

7. What has grief taught you? If you could tell other people who have not yet experienced grief one thing that you have learned about it, what would you tell them?

I suppose what I've learned, first and foremost, is that there is a difference between grief and grieving. Grieving is a process, a painful but necessary one all the same. Like all processes, it has a beginning, a middle, and (gratefully) an end. I am recovering from the grieving process—things have gotten better, easier, and I hope this trend continues.

Grieving changes a person, but lots of events in life change us. This is just another one. Grief, in contrast, is loss, and because that loss is permanent, so is grief. I don't think we ever "get over" grief. I will always miss my parents, Chris, Patty, and all the others I've lost in my life, and I will always mourn their absence. I've also learned that grief changes over time, which is a comforting thought. However, one of the more surprising things I've learned about grief is that some aspects of it are permanent. Not everything but certainly some things about grief will be with me for the rest of my life. I've come to see this as a consequence of the massive changes that losing a beloved partner creates in life. He is permanently gone, and the changes in me and in my life that his passing created are just as permanent. I will always miss him. I will always wish, however futile that wish might be, that he didn't have to be gone. But that yearning for him no longer brings everything else in my life to a halt. The more temporary aspects of grief include the bring-me-to-my-knees physical pain of that loss. In the beginning, I felt as though I'd lost the ability to see and appreciate happiness and joy in other people as well as in myself. That is not the case anymore. You might feel, at the outset, that you won't be able to laugh or smile again, but, at least for me, that is not what happens.

Grief changes us in many ways. I've come to understand that some of those changes have made me see myself and my life in a different light. There will be no return to the person I was and the life I had before this loss; time does not work that way. It is not possible to be profoundly changed by an experience and yet remain the same. Loss created a new normal for me, and that's the normal I go forward with. As Lamott and Rosemary say—I'm learning to dance with a limp.

It is, I think, often difficult to see the changes in ourselves when something like loss rips through our lives because they are changes we didn't and don't want, that we never asked for, and that affect almost everything about our lives. Our first response to changes that have been shoved down our throats, so to speak, is to push them away, to deny them, to refuse to deal with them until we have to. Adapting to those changes and incorporating them into our sense of who we are and where we're going happen in microsteps, often too small to see except in retrospect. With the passage of time, the new, changed person we are becomes normal. So, in speaking to my scared and anxious self at the time of his passing, I say, "Yes, you will feel 'normal' again—but not in the same way."

I love the answers my friends had to this question. They are not only brave women but also wise. S.R.H., whose loss is so recent, says: "Grief has taught me how to behave for others when their time comes. I am eternally grateful for the few people who have shown up to simply be there for me. To talk or not to talk. To bring groceries. To recount specific memories of something my husband said to them or did for them. To tell me why, in detail, they loved him, too."

From Edna, whose loss was so incredibly devastating and sudden, come words of hope and resilience: "You have to grieve in your own way, but try new things, so you don't get stuck in your grief. Everyone grieves differently. Don't be afraid of admitting you might need medical help and/or grief counseling. Reach out for help. Adopt a pet, join a club where you can meet new friends, and do things together."

Rosemary, Liz, and EC talk about how lucky they feel to have been able to share so many years with the person they loved, how grief has taught a deeper appreciation of even the small moments in life, and how to be more present in daily life. Catherine speaks of both the endlessness of grief and the gradual softening of its sharp edges: "It never gets better. That is not the way to phrase grief. It changes, it shifts, and the sharp edges of desperation and depression soften. Time allows you to establish a peaceful coexistence with it, but you carry it with you everywhere you go."

Sheila, thirty-nine years after the death of her husband, says that grief taught her to be "grateful for the time I had with my husband. . . . Many good memories. I can and do smile when I remember them." I'm not where Sheila is yet, but I hope to be.

On a bookshelf in my living room, I have a Daruma doll, a symbol of persistence and resilience in Japan. He is round, limbless, and bright red. The legend is that the Bodhidharma, the founder of Zen Buddhism, seeking enlightenment, silently sat without moving and meditated facing a wall for nine years. He

FIGURE 8.1. My Daruma doll, a Japanese symbol of patience, persistence, and resilience.

is depicted as a round ball, lacking arms and legs, because he was still and silent for so long his limbs fell off. He is red, the color of luck and good fortune. And he is round, with his weight at the bottom of the ball, toward the earth—un-tip-over-able, resilient, and able to bounce (or roll) back in the face of trouble. As I pass by that bookshelf, I occasionally pick him up, cup my hand around his bald head, and whisper, "Patience and resilience, little round guy," and then go on with my day. It makes me feel stronger, more patient, and more at peace. I will share him with you.

Peace to all of you.

Appendix

GRIEF STORIES

As you read these stories, I think you can see many of the aspects of grief and grieving I have explored in these chapters. The anger that many of the bereaved feel, directed at anyone and everyone, can easily be seen in both the women who are at the very beginning of the challenging path in front of them and the women who are decades beyond their loss. The physicality of the pain along with the psychological pain and stress of loss is evident in all of the stories told here. Even the permanence of the changes that loss creates, despite the softening of the sharp edges of grief over time, is plain in these stories. To everyone who replied to my intrusive questions, thank you. These replies have been edited only slightly.

In addition, because one of my major questions at the outset was about how long grief lasted, I've arranged the stories my friends tell in chronological order, beginning with the person who was closest to the loss (six months "out" from losing her husband) and ending with the person who had the most distance from that initial event. I can see the changes over time in the grief my friends were experiencing, and I hope you can, too.

SIX MONTHS

Written by S.R.H.

(a) *Whom did you lose?*

I lost my husband of nearly 54 years.

(b) *Can you tell me about the circumstances of that loss? How did your loved one die?*

He had gone into the hospital for a fairly routine prostate procedure and was expected to return home the same day. However, post-procedure, blood was not being flushed from his urine quickly enough, so the doctors debated for most of that day about whether or not to release him. At 11:00 p.m. they decided it would be best for him to stay the night. So he did. When I saw him the next day he was upbeat, funny, pain-free, and eager to come home. Nevertheless, they decided to keep him one more night. The following morning he called me very early, in terrible pain. I arrived at the hospital at the exact moment visiting hours began, 8:00, and he was still suffering dreadfully. At my insistent urging, the nurse finally gave him opium, which didn't take full effect for an hour. The plan was to transfer him to another room that morning, but as they prepared to do so he became dizzy, confused, and out of breath. Still, they went ahead with the transfer, and I trotted alongside his gurney. He pulled his COVID mask down to better breathe. He was panting, concentrating. His eyes became wild, his arm outstretched, then bent at the elbow toward his chest. We were searching each other's eyes when I heard a gurgle deep in his throat, and his eyelids fell to half-mast. I shouted to the transporter that something was very wrong and we needed help, so she wheeled him back to his original room, at which point the nurse called, "CODE!" and tens of people appeared out of nowhere.

Complete chaos. Two people ushered me into another room and closed the door to talk with me, but I said I needed to see what they were doing to my husband. They took me to him, holding me up, one on either side. I wasn't in danger of falling, but I allowed them to hold me. They had told me in the room that there would already be brain damage by this time, so I said the doctors should stop trying to save him, that he wouldn't want to be revived as less than himself. I told them to stop. "She said stop," someone said more loudly, so they did. They all filed out of the room, and I went to his side. I kissed his face and closed his eyes with my thumbs. I told him I loved him and asked him to come to me in my dreams tonight. They allowed me to stay with him for four hours, at which point his body was already cooling. It was both horrible and a gift to be with him at his moment of death. We were together until the end.

(c) *How long ago did the loss occur? How old were you when your loved one died?*

My husband died six months ago. I am 75 years old; he was three years older.

(d) *What is your opinion on the "medicalization" of grief? Should grief be treated like a disease?*

No, I don't think grief is a disease or should be treated like one. I'm not even sure what that would mean. It's part of the journey we will each of us travel, if we have been fortunate enough to have loved someone.

(e) *Do you see grief as having a purpose? If so, what is the purpose of grief?*

What is the purpose of grief? As a transition period, I suppose. From who you were to who you will become. Or maybe that's the purpose of grieving; I don't know.

(f) *Modern theories of grief and bereavement suggest that the pain and sorrow of grief are part of the sudden and usually unexpected nature of the loss of a loved one and part of our abrupt reorganization of our sense of self that the loss of that person creates. Do you think that your sense of self has changed? Are you a different person after the loss? What about you has changed?*

Of course I've changed. Profoundly. I feel diminished. I am a roller coaster of emotions. I, who like to think of myself as even-keeled, am now quick to anger. If I am with a group of people, I am out of synch with their conversation. If someone mentions his name, I think they have no right to say it so casually, but if they talk about what's going on in their lives, I wonder how they can be so callous. I am not myself. I don't know who I am.

(g) *What has grief taught you? If you could tell other people who have not yet experienced grief one thing that you have learned about it, what would you tell them?*

Grief has taught me how to behave for others when their time comes. I am eternally grateful for the few people who have shown up to simply be there for me. To talk or not to talk. To bring groceries. To recount specific memories of something my husband said to them or did for them. To tell me why, in detail, they loved him, too. Who do not say, "If I can do anything . . ." I cannot do your job for you—figure something out for yourself and do it. Or don't. But don't ask me to think.

ONE YEAR AND 117 DAYS

Written by Edna

(a) *Whom did you lose?*

I lost my husband, Thomas. We were high school sweethearts and had gone together for 2 years, before getting married in 1960.

Tom was only 18 and had a full-time job as a automobile mechanic after graduating from high school. I was 17 with a part time job at a dime store after graduating from high school a year after Tom. We were married 62 years. We had 2 children, 4 granddaughters, and 9 great-grandchildren.

b) Can you tell me about the circumstances of that loss? How did your loved one die?

Tom committed suicide. He was diagnosed approximately 5 years prior to his death with some form of dementia; however, he did not have Alzheimer's. His mother died almost 15 years after suffering from dementia, and his older brother Larry died after a lengthy bout of dementia as well. His brother was only 14 months older than Tom. Tom was fully aware of who everyone in our family was and where he was, unlike his mother and brother, but he knew he was beginning to lose his mind. Therefore, my son, Tommy, daughter, Debbie, and I think he just didn't want to become a burden and have to live in a nursing facility unable to take care of his own bodily functions.

c) How long ago did the loss occur? How old were you when your loved one died?

Tom has been gone a year and 117 days. He was 80 years old on the day he decided to kill himself, which happened to be our daughter Debbie's 60th birthday. I'm sure he never would have done this horrible thing on her birthday! If, he had been in his right mind, he never would have killed himself on her birthday. He loved his children and his grandkids so very much!! I was 79 years old at the time of his death.

(d) *What is your opinion on the "medicalization" of grief? Should grief be treated like a disease?*

Because my husband had terrible anger issues from the frustration of knowing he was suffering from dementia, there were times I was afraid he might physically hurt me in some way. On the day he killed himself, he had seemed fine and in fact happy. We had just arrived home when he rushed into the house, and as I started to enter the house, he opened the door and shot himself in the head in front of me. I don't think I really grasped what my life was or would be like for the first 2–3 months after his death. There were all kinds of legal things that I had to do following his death. Death! Where did that fit into my life. I just didn't comprehend that! I felt so guilty! Why didn't I know that Tom might be thinking of taking his own life? I was always so sad! Would I ever get over this feeling of guilt and loneliness! If it wasn't for my little dog Lizzie and knowing I had to take care of all her needs, I probably wouldn't have gotten out of bed. All I wanted to do was sleep. I felt so sick at my stomach all the time and started losing lots of weight. I couldn't sleep well and kept having nightmares. My son and daughter and granddaughters kept telling me I needed counseling and anxiety medication to help me get a handle on my grief. I knew they were very worried about me, and, being a mother, I certainly didn't want them worrying about me, especially since they were grieving as well. Two of my granddaughters are or were counselors and recommended a counselor to me. I finally called my doctor, who prescribed an anxiety medication, and then I went to see the counselor. After spending an hour with the counselor, she told me I needed this EMDR [Eye Movement Desensitization and Reprocessing] treatment. It is a treatment that she is certified to conduct in her office and is used to treat PTSD, which she said I had. The first treatment helped almost immediately, with the help of the medication as well. In my case, I wouldn't call it a disease, but I would call it a medical condition because I don't think I would have gotten better without the medical and counseling help I received.

(e) *Do you see grief as having a purpose? If so, what is the purpose of grief?*

I do believe that grief is the price for loving someone. I just wish there was a faster and more pain-free way of getting through my grief. There were times I felt like I was stuck. Like I could not move forward with my life. Fortunately, my children and family make me want to get through my grief, because I love them so much, and I don't want them to worry about me.

(f) *Modern theories of grief and bereavement suggest that the pain and sorrow of grief are part of the sudden and usually unexpected nature of the loss of a loved one and part of our abrupt reorganization of our sense of self that the loss of that person creates. Do you think that your sense of self has changed? Are you a different person after the loss? What about you has changed?*

I didn't want this change! I have had many changes in my life, and this is one of the hardest changes that I have had to go through. I never thought of myself as a widow. That was a very hard thing to have to write on various documents. I was always married before. I was my husband's partner. Now I am a single woman, a widow. Not something I ever wanted to be. My sense of self has definitely changed. I was always part of a couple. All our friends are couples. Now I find myself feeling like a fifth wheel. So I have started trying to make friends with other widows or single women. People I feel I have more in common with. Several of them are women in my neighborhood and senior organizations that I belong to. I'm going and doing things with new friends that Tom didn't want to do, like going on cruises and concerts and plays. Even though I still feel lonely, I know I need to add these new friends and things to my life to add more meaning to my life. In doing so, I feel like I am also helping my new friends as well.

(g) *What has grief taught you? If you could tell other people who have not yet experienced grief one thing that you have learned about it, what would you tell them?*

I would say, you have to grieve in your own way, but try new things, so you don't get stuck in your grief. Everyone grieves differently. Don't be afraid of admitting you might need medical help and/or grief counseling. Reach out for help. Adopt a pet, join a club where you can meet new friends, and do things together.

THREE YEARS

Written by Rosemary

(a) *Whom did you lose?*

I lost my husband, Tom.

(b) *Can you tell me about the circumstances of that loss? How did your loved one die?*

Tom died suddenly one Sunday afternoon. He seemed fine but said he felt tired and wanted to take a nap. I went to the living room to watch a movie. After about an hour, he called me, and I knew something was wrong. When I got to him, he said, "Call 911." He never said anything else. I believe he died right after that.

(c) *How long ago did the loss occur? How old were you when your loved one died?*

Tom died December 27, 2020. I was 63 at the time.

(d) *What is your opinion on the "medicalization" of grief? Should grief be treated like a disease?*

I was surprised by how physical my grief felt. My whole body reacted. I had never had high blood pressure, yet my blood pressure was high. I felt like my whole body was shaking. I asked my doctor for antidepressants. She explained that she did not believe that you could medicate your way out of grief. And I agree with this. Although for me grief was very physical (I felt sick), I agree that it isn't a disease. On the other hand, I don't believe that all of us will experience it at some point to the same degree. My husband never experienced the grief that I have. He lost parents and friends but never one that changed the essence of his life. I had lost my brother when I was a child (he was 18 and I was 10 when he was killed in a car accident). I remember the pain of that. It changed my whole life and that of my parents. But it didn't compare to losing a partner.

(e) *Do you see grief as having a purpose? If so, what is the purpose of grief?*

I wish I knew the purpose of grief. There are many different losses that cause it and perhaps different types of grief that result. At this time, I am focused on the loss of a partner. I agree that it is the price we pay for loving someone, and I wouldn't change anything about my relationship with Tom except I would have had him live forever. It's painful to think that one person out of every couple lucky enough to have a happy partnership must experience this grief.

(f) *Modern theories of grief and bereavement suggest that the pain and sorrow of grief are part of the sudden and usually unexpected nature of the loss of a loved one and part of our abrupt reorganization of our sense of self that the loss of that person creates. Do you think that your sense of self has changed? Are you a different person after the loss? What about you has changed?*

My sense of self has definitely changed. I saw myself as part of a couple, and I'm not sure that I realized it. I shared everything, joy and sadness, with someone. For example, during our marriage, I traveled a lot by myself. Yet part of my enjoyment was knowing that I would call Tom at home and share my trip.

Although I can still share my experiences with my children and friends, it is not the same. Travel and so many other things are not the same now. As Anne Lamott wrote, "It's like having a broken leg that never heals perfectly—that still hurts when the weather gets cold, but you learn to dance with the limp." I'm still learning to dance again.

(g) *What has grief taught you? If you could tell other people who have not yet experienced grief one thing that you have learned about it, what would you tell them?*

I'm not sure what grief has taught me other than how lucky I was to have shared a life with Tom. Now I am trying to build a new life with him as only a memory, and it is not fun. From the death of my brother, I learned to treasure every moment, and I never lost the feeling that you can lose everything in a moment. I'm not sure everyone understands that, but I don't know if you truly can understand without having experienced such loss. I would tell people who are grieving to seek out others who are also grieving or who have experienced such grief. I find comfort in sharing stories of grief.

SEVEN YEARS

Written by Liz

(a) *Whom did you lose?*

Charles and I met on an epic blind date, set up by mutual friends. Though "opposites," we were best friends for 32 years and married for 29 years. While we had some typical "bumps in the road" over three decades together, he was a wonderful father, partner, and storyteller.

(b) *Can you tell me about the circumstances of that loss? How did your loved one die?*

I eventually learned the cause was an aortic aneurysm, something with a 5 percent chance of survival even if you are in a hospital when it occurs. The medical examiner's office indicated that it was almost instantaneous, as Charles would have wished and a small comfort to our son and me. I do regret that he did not go speak to a cardiologist when he occasionally referenced some chest pain twinges after a long workout, but grown men make up their own minds about health care.

Charles had been a football player in high school, and a number of his teammates died young of heart attacks. We had talked about it, and I could tell it was in the back of his mind in our '50s. He was a big guy physically who worked out at the gym daily for 90 minutes or more and ate cleanly, so I never worried that he might die young. Still, he'd expressed a wish to go quickly when it was his time, and I knew he wanted to be cremated, but those were the only points related to dying that we had discussed.

I work for a small college and had to accompany a group of students late one October Wednesday night. I called Charles about 6:00 p.m., and we talked about our days. I reminded him of my obligation that evening, I told him to eat dinner without me, that I'd be home about 9:00 or 10:00 p.m., and we said we loved each other.

During the event with the students, about 8:00 p.m., I got a strong feeling that something was very wrong. I wanted to leave but knew the students were counting on me to stay, so I did. When the event wrapped up, I drove home as fast as I could. I got there about 10:00 p.m. and parked on the street at the bottom of our hill. It struck me as odd that the trash barrel was not out for collection the next morning and the porch light was off. Charles always took care of those things. I hurried up the hill to the front door, and as soon as I opened it, I could see Charles at the back of the house, in "his chair" in the den, not looking right. He did not answer when I called his name. When I got to him, he was cool to the touch and had a bluish pallor. His laptop was on the side table next to him, and he looked at peace, as if he was working one minute and gone the next.

I immediately called 911, and the dispatcher wanted me to pull him onto the floor and start chest compressions. I knew that would be futile; I had irrational thoughts of banging his head on the hard wood and causing him brain damage; plus, he was bigger than I was, so it was difficult to move him. Yet I did, and then I waited on the first responders. Our street was narrow with houses close together. It surprised me that all the sirens didn't raise my neighbors' curiosity. Friends on either side of us slept through it. I had to notify them the next day and to deal with a neighborhood Facebook thread: "I hope nothing bad happened last night on X Street. What was up with all the emergency vehicles?"

First, a police officer arrived and asked to see my driver's license to verify that I lived there and was who I said I was. He was fairly young, thirtyish, and I think it may have been the first time he'd had a call with a deceased person. I was in shock and not crying, which seems peculiar looking back; it was as if I was observing all of this from above. I sat on the living room couch with the cop across the room, trying to process what to do next and to not look back at Charles, undignified on the floor 50 feet away. Next, the EMTs came, pronounced Charles dead, and quickly left after calling the coroner's office. That team soon arrived, said an autopsy would be performed, and they took Charles away. I had the presence of mind to remove his wedding band. It was about 11:30 p.m. at that point, and the police officer did not want to leave me until someone else arrived to be with me.

I called my dad across town, but his cell phone was off, so I left a message to call me in the morning. Next, I called my mother-in-law, who lived 4 hours south of us; she immediately knew something was amiss given the lateness of the hour and the fact that normally Charles, not me, would be initiating a call. Her keening I will never forget, as Charles was her first-born, much loved, and he had recently been visiting her every month to help her with my father-in-law, who was battling dementia. (The police officer was a silent witness to this drama playing out.) Next, I called our recently graduated from college only son whom Charles had helped move across the country to Texas just two weeks prior. That was a horrible call, not just because of the news I was delivering but because I knew he had no one nearby to lean on. His cries, disbelief, and wailing also stick with me to today; he and his father were extremely close. Once he could speak, we agreed that he would book a flight home early the next morning. My last call was to my middle brother, who is great in a crisis. He was 2.5 hours away but insisted on jumping in his car immediately and driving to my house after calling our younger brother to give him the news. Once the cop heard that an adult was on their way to me, he agreed to leave. I know he meant well; it just felt intrusive at the time, although I later took homemade brownies to his precinct to thank him, and he dropped by on a rainy night 3 months later to check on me.

(c) *How long ago did the loss occur? How old were you when your loved one died?*

Charles's "angelversary" was 6 years ago last month. I was 53, almost 54, and he was 59, about to turn 60. October is a difficult month each year. Halloween can be very triggering for widowed people. My mother died (at age 64) on October 4, Charles died on October 5, his birthday was October 12, and his parents' birthdays were later that month. We put off his memorial until October 21 to give ourselves some time to plan and to grapple with the shock; that day was actually comforting and somewhat like a 60th birthday party for Charles.

(d) *What is your opinion on the "medicalization" of grief? Should grief be treated like a disease?*

I do not consider grief a disease, but it certainly can have biological symptoms, such as "widow's fog" for the first year (like "pregnancy brain"—you are forgetful and need to write down important things or create to-do lists). Obviously, sleeping and eating patterns can be impacted. Fear can be an issue. Anger too. The latter two caused me problems in the first and second years out, respectively.

(e) *Do you see grief as having a purpose? If so, what is the purpose of grief?*

I see love as a continuum. Love and grief are experienced differently for "your person" compared to a grandparent, parent, friend, or pet's death. We had a deep love, and while the ache has dissipated some now, I will always feel a hole in my life because Charles was a huge part of it for more than 30 years. And he's my son's dad, and my family still talks about him.

I do think grief has a purpose; you must work through its stages as you reframe *everything* about your life. And everyone's grief timeline is unique. I hate telling newly widowed folks that year 2 is worse than year 1, but it's true. Reality sets in after the first year of "busyness" with estate settlement, canceling accounts, perhaps moving (as I did—since it was traumatic to come home to the image of finding him gone each night, I sold that house after 6 months). Also, people initially cluster around, then subsequently fade into the background as months roll along. I say it "doesn't get better, it gets different." The United States does not handle death well as a society. People can utter the stupidest things even if they mean to be helpful ("He's in a better place." "You're young, you'll find someone else." "Call me if I can do anything" when they are the ones who should call or text you and offer X, Y, or Z.)

(f) *Modern theories of grief and bereavement suggest that the pain and sorrow of grief are part of the sudden and usually unexpected nature of the loss of a loved one and part of our abrupt reorganization of our sense of self that the loss of that person creates. Do you think that your sense of self has changed? Are you a different person after the loss? What about you has changed?*

Yes, it was a struggle, but my sense of self definitely has changed. I joke (half-heartedly) that I have become the "designated widow" on my college campus. If a trustee or employee's friend or family member loses their spouse, I get an email asking that I reach out with support (which I am glad to do). It was sad that in quick succession after my own loss, four faculty friends became widows, my father-in-law died three months after Charles, and my former sister-in-law lost her second husband to cancer at 40.

I had to find new ways to fill up my non-working hours; my father used to fuss that I did too much volunteer work, but it was a defense mechanism. I was no longer part of a couple (and couple friends stop inviting you to events, as if death is contagious). I had to learn to cope with being alone in my new home. I threw myself into decorating it in my own style, which was a good thing, following the long search for something I wanted to and could afford to buy, near work, and 30 minutes away from our former neighborhood. I do miss my former neighbors, except for the one that "knew I'd be selling," so she gleefully waited for a for-sale sign to pop up in my yard. As a longtime renter, it offered her a chance to permanently move onto our street. I confess that I was relieved to have higher bids to consider and did not choose to sell to her. After five years in my townhouse, I have acclimated to my current part of town and new neighbors there. I believe that Charles would want me to be living fully for both of us.

I resisted for the first year or so when my son tried to talk me into seeking out a grief therapist to help me deal with non-specific anger (at being widowed, at Charles's life being cut short when we were planning to start traveling, at having a much tighter budget since single people are taxed more than couples and Charles died without adequate life insurance).

The therapist was a true blessing. I saw her weekly for over a year and am so grateful. Our conversations helped me get over being "scared" generally—a hard thing since I was pretty "fearless" before Charles died. I had started to consider dating again in year 3, and she was wise in helping me sort through whether I loved myself enough in a healthy way to consider launching a new relationship. There certainly were some rough learning experiences with scammers and scoundrels to move past—always a frightening thing for (widowed) people of any gender and unfortunately very common with online dating. I went through all the dating apps, learned to not take every rejection personally, to take breaks when the "swipe syndrome" got to be too much. Happily going into year 7, I now have found someone worthy of my time who treats me well and is fun to spend time with—which is perfect for now even if I don't remarry. Certainly, not every widow/er will want to date again; it's a personal decision. I liken it to loving more than one child; your heart has room for more than one, and the right person will not be threatened by your past. I am not the same person I was when Charles and I met on that first blind date or that I was at 53. Friends say, "You are so strong," but most widowed people will tell you we don't really have a choice in that.

I also found a social support group for widowed people (Soaring Spirits International) about a year in. I highly recommend attending one of their "Camp Widow" weekends. That organization has been a lifeline, to the point that the local chapter leader/founder stepped down and asked me to take over two years ago. I no longer feel as if I need to talk as often about my grief; it helps me to provide a model of stability, of getting to the point where one can laugh and smile and have hope again, especially for the many young wids I've encountered in their 20s and 30s. One should not compare life paths, but it's hard not to be thankful that I had Charles for 32 years.

(g) *What has grief taught you? If you could tell other people who have not yet experienced grief one thing that you have learned about it, what would you tell them?*

The first question is addressed above.

I do tell family and friends who have not been widowed to appreciate the time they have with their "person," to work on any disagreements or to make a change—life is short. Buy life insurance. Take that trip! Don't put things off. And it's okay to tell a widow/er that you don't know what they are going through, but you're willing to listen or just sit quietly with them, to go for coffee or a walk, or to speak about their lost person by name (we do not want them to be forgotten).

TEN YEARS

Written by EC

(a) *Whom did you lose?*

I lost my husband of 30 years.

(b) *Can you tell me about the circumstances of that loss? How did your loved one die?*

He died 2½ months after being diagnosed with pancreatic cancer. In the summer before he died, he finished the Appalachian Trail (from Springer Mountain Georgia to Katahdin Maine).

It took him seven years to hike the entire trail in sections. After he succeeded, we hosted a big party in which he presented a slide show and celebrated this achievement. Soon after he developed an unexplained abdominal pain and after a scan was diagnosed. We went to MD Anderson (to the doctor that was featured in the book *The Last Lecture*), and he started a newer version of chemotherapy that may have given us a year more. However, he died sooner than expected.

(c) *How long ago did the loss occur? How old were you when your loved one died?*

The loss occurred in 2013. I was 62 at the time. My children were grown and living away from home. We were looking forward to more adventures. Fortunately, we did spend three weeks in Italy the spring before he died.

(d) *What is your opinion on the "medicalization" of grief? Should grief be treated like a disease?*

I think grief "symptoms" vary with the individual, the specific relationship, and the life stage of the people involved. I think grief can overlap with significant depression and anxiety or can create those strong feelings in the person left behind. Although

"Prolonged Grief Disorder" has recently been identified, I believe this diagnosis can be inappropriately applied for a natural, very painful experience that disrupts the major attachment of our lives. When others try to specify how long the grief "should" last, I think this compounds the grief.

(e) *Do you see grief as having a purpose? If so, what is the purpose of grief?*

I'm not sure it has a purpose, but it is a reality of life that we, at least many of us in the US, are unprepared for. When we get partnered, we are looking ahead to time together, not anticipating the loss. Looking back now, it seems I should have considered this possibility more seriously a long time ago. My husband and I should have talked about this possible outcome. I don't know if that would have helped, but it may reduce the shock of it all.

(f) *Modern theories of grief and bereavement suggest that the pain and sorrow of grief are part of the sudden and usually unexpected nature of the loss of a loved one and part of our abrupt reorganization of our sense of self that the loss of that person creates. Do you think that your sense of self has changed? Are you a different person after the loss? What about you has changed?*

I do think I've changed as a result of this disruption to my life. My sense of security and connection have been challenged, and I have had to reorganize my view of myself. I've experienced intense loneliness and a sense of being disconnected. Over time, these negative feelings have lessened, and I found myself dividing up my needs by forging stronger attachments to a few close friends and relatives rather than having them primarily met in my spouse.

(g) *What has grief taught you? If you could tell other people who have not yet experienced grief one thing that you have learned about it, what would you tell them?*

I've learned to be more present in my day-to-day life, although I still struggle to maintain such a mindful approach. I appreciate the smaller things more and remind myself more often to notice the good things in my life.

At the same time, I've had an increase in anxiety about losing others in my life. I recently had one of my closest friends die. I sometimes feel a cumulative impact of sequential losses and start to anticipate more of these adjustments.

I believe I am more understanding of loss and respond better to the losses of others. I now show up (to the funeral, to the home, etc.) more readily without wondering if this is the right thing to do. I know it is.

TWENTY-FIVE YEARS

Written by Catherine

(a) *Whom did you lose?*

My husband of 12 years. We were married from 1985 until 1997.

(b) *Can you tell me about the circumstances of that loss? How did your loved one die?*

We were on July 4th (1997) vacation at my mom's house in Hilton Head, SC. It was July 5, the day after, and we were planning another trip to the beach. I went to the grocery store to grab a bag of ice, and when I came back, I asked Mom, "Where's Jeff? He said he would be ready to go." She said she hadn't heard anything. I walked back to our room/my room, and he was face down on the floor. I called 911, and the operator walked me through CPR, but I knew he was dead. After the EMT people arrived, I kept asking/telling them that he was dead, but they kept saying that they weren't allowed to confirm deaths.

We followed them to the hospital and got the confirmation of death within 15 minutes of our arrival.

(c) *How long ago did the loss occur? How old were you when your loved one died?*

Twenty-five years ago. I was 39. He was 41.

(d) *What is your opinion on the "medicalization" of grief? Should grief be treated like a disease?*

I do think that aspects of grief have been medicalized, especially the frequent references to PTSD in my case, as well as the quick and easy way that the psychiatrists and my primary care doctor prescribed Xanax and various antidepressants. I found that I got more out of visits to psychologists and grief support groups than anything else. I actually think that grief and grieving are the opposite of disease in many respects, although I also guess it depends upon what kind of disease you're talking about. Quarantining, for example, is usually the opposite of what a grieving person wants. I wanted to talk about Jeff, look at pictures, tell stories, and talk to our friends. Diseases usually have a beginning and an end, or at least a remission. Grief stays with you; it transforms into various modes, and it never really gets better. It changes over time, of course, but it can manifest in the oddest moments, and it stays with you forever. That's also why I never bought the stages-of-grief author's shock, anger, disbelief, etc. It's too neat. I can still feel utter shock and anger today, depending on what's happening.

(e) *Do you see grief as having a purpose? If so, what is the purpose of grief?*

At one of my grief support group meetings, someone said that one day instead of saying, "Jeff died," I will be able to say, "Jeff lived." I have been trying to say that ever since and still can't. That's because I can't see how dying suddenly of a heart attack at age 41 could ever make sense. It has no purpose. And because the death had no purpose, I can't see how the paralyzing despair, anger, fear of dying, fear of everyone around me dying could have purpose other than pushing me down so far into a deep dark hole that I didn't think I could ever make it out alive.

(f) *Modern theories of grief and bereavement suggest that the pain and sorrow of grief are part of the sudden and usually unexpected nature of the loss of a loved one and part of our abrupt reorganization of our sense of self that the loss of that person creates. Do you think that your sense of self has changed? Are you a different person after the loss? What about you has changed?*

I do think that my sense of self has changed, but not in big sweeping ways. I empathize with people who suddenly lose a loved one. I can say with some authority that I know what that is like. At the same time, I have become cynical about death. It always seems to be around, hovering, waiting to snatch someone and create unspeakable devastation. I think I have also built stamina for subsequent disappointments, sadness, aimlessness, anger, and all the other emotions associated with death. I think I can actually say there are worse things. I think it's made me stronger in that sense.

(g) *What has grief taught you? If you could tell other people who have not yet experienced grief one thing that you have learned about it, what would you tell them?*

It never gets better. That is not the way to phrase grief. It changes, it shifts, and the sharp edges of desperation and depression soften. Time allows you to establish a peaceful coexistence with it, but you carry it with you everywhere you go. Is that a good thing? I wish I could go back and live life for the past 25 years without it, and then I could answer this with more certainty. It is an unmovable object that can present itself in myriad ways.

TWENTY-EIGHT YEARS

Written by LC74

(a) *Whom did you lose?*

We all face losses in our life, and sometimes the circle of life gets broken in a way we never expected. The hardest and most difficult loss to accept or recover from has been the loss of my youngest son at the age of 15. Twenty-eight years later that acceptance or recovery has still not been achieved, and throughout the years it takes on different levels of pain. Birthdays, holidays, and family events bring back the memories of those early childhood days and the sorrows of not being able to share in watching him grow up and have a family of his own.

(b) *Can you tell me about the circumstances of that loss? How did your loved one die?*

One never knows when your life can be totally turned upside down and forever altered. The simple act of waking the kids up in the morning for school in the same way as always with a simple kiss on the cheek and a good morning, sending them out the door with a "see you after school" all came crashing down with a call from the Trauma Center Hospital with a urgent message to come immediately as your son has been brought in after a serious car accident. Three days later the horrible decision had to be made to take him off life support. That decision will continue to haunt me for the remainder of my life.

(c) *How long ago did the loss occur? How old were you when your loved one died?*

Twenty-eight years seems like only yesterday as so many details of those days before the accident and the days following float through your head at night when sleep fails you. Years, months, days, minutes, seconds have no measure in dealing with the loss of your son.

(d) *What is your opinion on the "medicalization" of grief? Should grief be treated like a disease?*

Grief cannot be defined as a single medical condition, nor can it be treated with a single medical prognosis or treatment. That could be compared to someone treating cancer with the same treatment for every cancer that exists. We all know that cancer comes in many forms, and we have all dealt with the loss of a friend or family member from that hideous disease, and yet we know that there is no one "medical" treatment for that disease. In the same fashion, how grief is manifested in each of our lives is different, and some may look to medical professionals to deal with the sorrow and pain, and others will deal with it on a personal fashion. In my case I've dealt with the "mental condition" by getting out on the golf course or out on the water in a kayak . . . spending time alone to remember the happy moments and never to forget those times is the best treatment. Having recently added to the "grief burden" the loss of my husband to cancer, I've found myself doing those same things again and now add to those memories.

(d) *Do you see grief as having a purpose? If so, what is the purpose of grief?*

No, I don't believe there is a purpose of grief; rather, one learns how to deal with grief, and it is our purpose to share with others going through similar circumstances and help them to understand what has happened in their life and offer support and care. I've seen so many circumstances of people that have not experienced loss that say or do things that are so inappropriate to the individual facing a difficult time in their life. It is my goal to share with them ways to lift up [our] care in those impossible periods.

(f) *Modern theories of grief and bereavement suggest that the pain and sorrow of grief are part of the sudden and usually unexpected nature of the loss of a loved one and part of our abrupt reorganization of our sense of self that the loss of that person creates. Do you think that your sense of self has changed? Are you a different person after the loss? What about you has changed?*

Certainly, those unexpected losses lead to an altering of your purpose in life, your personality, your relationships, and everything surrounding you. This loss led to a different focus, involvement in volunteer activities, (a divorce 5 years after his death as we both dealt with the loss in a different fashion and drifted apart), and now in retirement after the loss of my son and second husband, I am determined to share as much time with my other sons and grandchildren, taking them on memorable trips and activities and in general taking that independence now to focus on myself and travel and do things for myself.

THIRTY-NINE YEARS

Written by Sheila

(a) *Whom did you lose?*

My husband

(b) *Can you tell me about the circumstances of that loss? How did your loved one die?*

He died from cancer.

(c) *How long ago did the loss occur? How old were you when your loved one died?*

[He] died in November of 1984. I was 47.

(d) *What is your opinion on the "medicalization" of grief? Should grief be treated like a disease?*

It causes pain, and perhaps in that way [it] is like a serious illness.

(e) *Do you see grief as having a purpose? If so, what is the purpose of grief?*

Perhaps true in that it later becomes fond memories and gratitude for the time we had.

(f) *Modern theories of grief and bereavement suggest that the pain and sorrow of grief are part of the sudden and usually unexpected nature of the loss of a loved one and part of our abrupt reorganization of our sense of self that the loss of that person creates. Do you think that your sense of self has changed? Are you a different person after the loss? What about you has changed?*

That in a way you are fortunate. Being loved and loving are treasures. In that way, you have been fortunate.

(g) *If you could tell other people who have not yet experienced grief one thing that you have learned about it, what would you tell them?*

I was grateful for the time I had with my husband, Foster. Many good memories. I can and do smile when I remember them.

NOTES

1. THE BEGINNING OF GRIEF

1. U.S. Institute of Medicine, Committee for the Study of Health Consequences of the Stress of Bereavement, *Bereavement: Reactions, Consequences, and Care*, ed. Marian Osterweis, Frederic Solomon, and Morris Green (National Academy Press, 1984), chap. 4: "Reactions to Particular Types of Bereavement," full text available online at https://www.ncbi.nlm.nih.gov/books/NBK217842/pdf/Bookshelf_NBK217842.pdf.
2. Lisa Shulman, *Before and After Loss: A Neurologist's Perspective on Loss, Grief, and Our Brain* (Johns Hopkins University Press, 2018), 10.
3. The Doors, "Five to One," track 5, side 2, on *Waiting for the Sun* (Elektra, 1968).
4. World Health Organization, WHO COVID-19 Dashboard, https://data.who.int/dashboards/covid19/cases?n=c, accessed October 2023.
5. Arielle Schwartz, "Grief, Grit, and Grace," *Dr. Arielle Schwartz, PhD* (blog), Center for Resilience Informed Therapy, March 8, 2016, https://drarielleschwartz.com/grief-grit-and-grace-dr-arielle-schwartz/#.Y5zN53bMJD8.
6. Avril Maddrell, "Mapping Grief: A Conceptual Framework for Understanding the Spatial Dimensions of Bereavement, Mourning, and Remembrance," *Social and Cultural Geography* 17, no. 2 (2016): 166–88, https://dx.doi.org/10.1080/14649365.2015.1075579; "Grief vs. Grieving: What Is the Difference?," *Grieveleave* (blog), February 12,

2023, https://www.grieveleave.com/blog/griefvsgrieving#:~:text=Grief%20is%20the%20collection%20of,or%20thing%20you've%20lost.

7. Sidney Zisook and Katherine Shear, "Grief and Bereavement: What Psychiatrists Need to Know," *World Psychiatry* 8 (2009): 67.
8. John Archer, *The Nature of Grief: The Evolution and Psychology of Reaction to Loss* (Routledge, 1999), 66–91.
9. Ivette Hidalgo, Dorothy Brooten, JoAnne M. Youngblut, Rosa Roche, Juanjuan Li, and Ann Marie Hinds, "Practices Following the Death of a Loved One Reported from 14 Countries or Cultural/Ethnic Group," *Nursing Open* 8 (2021): 453–62, https://doi.org/10.1002/nop2.646.
10. Caroline Lloyd, *Grief Demystified: An Introduction* (Jessica Kingsley, 2018), 24.
11. Kelly Diane Mead, "Transcending Culture: The Universality of Grief" (master's thesis, Smith College, 2007), 1, https://scholarworks.smit.educ/theses/1346.
12. Lloyd, *Grief Demystified*, 24–25.
13. Paul C. Rosenblatt, "Grief in Small Scale Societies," in *Death and Bereavement Across Cultures*, 2nd ed., ed. Colin Murray Parkes, Pittu Laungani, and Bill Young (Routledge, 2015), 25.
14. These other forms are described in PDQ Supportive and Palliative Care Editorial Board, "Grief, Bereavement, and Coping with Loss (PDQ®)," health professional version, National Cancer Institute PDQ Cancer Information Summaries, June 26, 2024, https://www.ncbi.nlm.nih.gov/books/NBK66052/.
15. Shulman, *Before and After Loss*, 1.
16. Lynn Eldridge, "How Anticipatory Grief Differs from Grief After Death," Verywell Health, Health A–Z, updated July 15, 2023, https://www.verywellhealth.com/understanding-anticipatory-grief-and-symptoms-2248855.
17. Holly G. Prigerson, Mardi J. Horowitz, Selby C. Jacobs, Colin M. Parkes, Mihala Aslan, Karl Goodkin, et al., "Prolonged Grief Disorder: Psychometric Validation of Criteria Proposed for DSMV and ICD-11," *PLoS Med* 6, no. 8 (2009): art. e1000121, https://doi.org/10.1371/journal.pmed.1000121.

18. Kelley Lynn, "I Will Never Move On," *Rip the Life I Knew* (blog), February 18, 2016, capitalization in original, reposted on Soaring Spirits International, Widow's Voice, February 19, 2016, https://widowsvoice.com/post/i_will_never_move_on/.
19. John Bowlby, *Attachment and Loss*, 3 vols. (1969; reprint, Basic, 1982); Archer, *The Nature of Grief.*
20. Claire White and Daniel M. T. Fessler, "Evolutionizing Grief: Viewing Photographs of the Deceased Predicts the Misattribution of Ambiguous Stimuli by the Bereaved," *Evolutionary Psychology* 11, no. 15 (2013): 1085.
21. Randolph M. Neese, "An Evolutionary Framework for Understanding Grief," in *Spousal Bereavement in Late Life*, ed. Deborah Carr, Randolph M. Neese, and Camille B. Wortmann (Springer, 2006), 205.
22. White and Fessler, "Evolutionizing Grief."
23. Archer, *The Nature of Grief.*
24. Miguel Farias, Anna-Kaisa Newheiser, Guy Kahane, and Zoe de Toledo, "Scientific Faith: Belief in Science Increases in the Face of Stress and Existential Anxiety," *Journal of Experimental Social Psychology* 49 (2013): 1210–13; Aaron C. Kay, Jennifer A. Whitson, Danielle Gaucher, and Adam D. Galinsky, "Compensatory Control: Achieving Order Through the Mind, Our Institutions, and the Heavens," *Current Directions in Psychological Science* 18 (2009): 264–68.
25. Gang Chen, B. Douglas Ward, Stacy A. Claesges, Shi-Jiang Li, and Joseph S. Goveas, "Amygdala Functional Connectivity Features in Grief: A Pilot Longitudinal Study," *American Journal of Geriatric Psychiatry* 28, no. 10 (2020): 1089–101, https://doi.org/10.1016/j.jagp.2020.02.014.

2. LOVE, ATTACHMENT, AND GRIEF

1. Lisa Shulman, *Before and After Loss: A Neurologist's Perspective on Loss, Grief, and Our Brain* (Johns Hopkins University Press, 2018), 58.
2. Shulman, *Before and After Loss*, 59.
3. Quoted in Rick Marshall, "Obscenity Case Files: *Jacobellis v. Ohio* ('I Know It When I See It')," CBLDF (Comic Book Legal Defense Fund), https://cbldf.org/about-us/case-files/obscenity-case-files/, accessed March 3, 2025.

4. *APA Dictionary of Psychology*, s.v. "love, *n.*," updated November 15, 2023, https://dictionary.apa.org/love.
5. Elaine Hatfield and Richard L. Rapson, "Love and Attachment Processes," in *Handbook of Emotions*, 2nd ed., ed. Michael Lewis and Jeannette M. Haviland-Jones (Guilford Press, 1993), 654–62.
6. Robert Plutchik, "The Nature of Emotions: Human Emotions Have Deep Evolutionary Roots, a Fact That May Explain Their Complexity and Provide Tools for Clinical Practice," *American Scientist* 89, no. 4 (2001): 345, 346.
7. Hokuma Karimova, "The Emotion Wheel: What It Is and How to Use It," Positive Psychology.com, December 24, 2017, https://positivepsychology.com/emotion-wheel/.
8. John Bowlby, *Attachment*, vol. 1 of *Attachment and Loss* (1969; reprint, Basic, 1982), 198–209.
9. Bowlby, *Attachment*, 209.
10. John Bowlby, *Loss, Sadness, and Depression*, vol. 3 of *Attachment and Loss* (1969; reprint, Basic, 1982), 7.
11. Myron Hofer, "Relationships as Regulators: A Psychobiological Perspective on Bereavement," *Psychosomatic Medicine* 46, no. 3 (1984): 183–97.
12. C. S. Lewis, *A Grief Observed* (1961; reprint, New York: Harper Collins, 1996), 47.
13. Hofer, "Relationships as Regulators," 184.
14. Selby C. Jacobs, Thomas R. Kosten, Stanislav V. Kasl, Adrian M. Ostfeld, Lisa Berkman, and Peter Charpentier, "Attachment Theory and Multiple Dimensions of Grief," *Journal of Death and Dying* 18, no. 1 (1988): 41–52, https://doi.org/10.2190/8QD0-W9R6-QX96-A5; Lisa M. Diamond and Janna A. Dickenson, "The Neuroimaging of Love and Desire: Review and Future Directions," *Clinical Neuropsychiatry: Journal of Treatment Evaluation* 9, no. 1 (2012): 39–46.
15. Phillipa Lally, Cornelia, H. M. van Jaarsveld, Henry W. W. Potts, and Jane Wardle, "How Are Habits Formed: Modelling Habit Formation in the Real World," *European Journal of Social Psychology* 40 (2010): 998–1009.
16. "How a Secure Attachment Style Develops in Early Childhood," The Attachment Project, May 21, 2021, updated October 3, 2024,

https://www.attachmentproject.com/blog/secure-attachment-style-in-early-childhood/.

17. Elisabeth Kubler-Ross, *On Death and Dying* (Routledge, 1969); Sigmund Freud, "Mourning and Melancholia" (1917), in *The Standard Edition of the Complete Psychological Works of Sigmund Freud*, vol. 14: *1914–1916: On the History of the Psycho-Analytic Movement, Papers on Metapsychology, and Other Works*, trans. and ed. James Strachey and Anna Freud (1967; reprint, Hogarth Press, 1986), 243–58.
18. Margaret Stroebe and Henk Schut, "The Dual Process Model of Coping with Bereavement: A Decade On," *Omega* 61, no. 4 (2010): 197–224.
19. Stephanie Ortigue, Francesco Bianchi-Demischeli, Nisa Patel, Chris Frum, and James W. Lewis, "Neuroimaging of Love: fMRI Meta-analysis Evidence Toward New Perspectives in Sexual Medicine," *Journal of Sexual Medicine* 7 (2010): 3541–52, https://doi.org/10.1111/j.1743-6109.2010.01999.
20. Claudio Lavin, Camilo Melis, Ezequiel Mikulan, Carlos Gelormini, David Huepe, and Agustin Ibanez, "The Anterior Cingulate Cortex: An Integrative Hub for Human Socially-Driven Interactions," *Frontiers in Neuroscience* 7 (2013): art. 64, https://doi.org/10.3389/fnins.2013.00064.
21. Harald Gundel, Mary-Frances O'Connor, Lindsey Littrell, Carolyn Fort, and Richard D. Lane, "Functional Neuroanatomy of Grief: An fMRI Study," *American Journal of Psychiatry* 160 (2003): 1946–53, https://doi.org/10.1176/appi.ajp.160.11.1946.
22. Robert Leech, Rodrigo Braga, and David J. Sharp, "Echoes of the Brain Within the Posterior Cingulate Cortex," *Journal of Neuroscience* 32, no. 1 (2012): 215–22, https://doi.org/10.1523/JNEUROSCI.3689-11.2012.

3. GRIEF AND STRESS: THE PHYSIOLOGICAL EFFECTS OF BEREAVEMENT

1. U.S. Institute of Medicine, Committee for the Study of Health Consequences of the Stress of Bereavement, *Bereavement Reactions, Consequences, and Care*, ed. Marian Osterweis, Fredric Soloman, and Morris Green (National Academies Press, 1984), 163, full text available

online at https://www.ncbi.nlm.nih.gov/books/NBK217842/pdf/Bookshelf_NBK217842.pdf.

2. Myron Hofer, "Relationships as Regulators: A Psychobiological Perspective on Bereavement," *Psychosomatic Medicine* 46, no. 3 (1984): 193.
3. Hans Selye, "The General Adaptation Syndrome and the Diseases of Adaptation," *Journal of Clinical Endocrinology and Metabolism* 6, no. 2 (1946): 117–230.
4. James L. McGaugh, "Making Lasting Memories: Remembering the Significant," *Proceedings of the National Academy of Sciences* 110 (suppl. 2) (2013): 10402–7; Juliana Nery de Souza-Talarico, Marie-France Marin, Shireen Sindi, and Sonia J. Lupien, "Effects of Stress Hormones on the Brain and Cognition: Evidence from Normal to Pathological Aging," *Dementia and Neuropsychologica* 5, no. 1 (2011): 8–16; Gregory E. Miller, Edith Chen, and Eric S. Zhou, "If It Goes Up, Must It Come Down? Chronic Stress and the Hypothalamic-Pituitary-Adrenocortical Axis in Humans," *Psychological Bulletin* 133, no. 1 (2007): 25–45.
5. Selye, "The General Adaptation Syndrome and the Diseases of Adaptation."
6. Margaret Stroebe, Henk Schut, and Wolfgang Stroebe, "Health Outcomes of Bereavement." *The Lancet* 370, no. 9603 (2007): 1960–73.

4. THE EMOTIONS OF GRIEF

1. Robert W. Levenson, "What Is the Function of Emotion?," in *The Nature of Emotion: Fundamental Questions*, ed. Paul Ekman and Richard J. Davidson (Oxford University Press, 1994), 123–26.
2. Chris Isaak, "Don't Leave Me on My Own," track 5 on *Forever Blue*, CD (Studio D and Dave Wellhausen Recording, 1995).
3. Guy Winch, "The Important Difference Between Sadness and Depression and Why so Many Get It Wrong," *Psychology Today*, October 2, 2015, https://www.psychologytoday.com/us/blog/the-squeaky-wheel/201510/the-important-difference-between-sadness-and-depression.
4. Bonanno paraphrased in Hara Estroff Marano, "At a Loss," *Psychology Today*, June 19, 2020, https://www.psychologytoday.com/us/articles/202006/loss.

5. Joseph P. Forgas, "Can Sadness Be Good for You? On the Cognitive, Motivational, and Interpersonal Benefits of Mild, Negative Affect," *Australian Psychologist* 52 (2017): 3–13.
6. Galen V. Bodenhausen, Lori A. Sheppard, and Geoffrey P. Kramer, "Negative Affect and Social Judgment: The Differential Impact of Anger and Sadness," *European Journal of Social Psychology* 24 (1994): 45–62.
7. Joseph P. Forgas, "Four Ways Sadness May Be Good for You," *Greater Good Magazine*, June 4, 2014, https://greatergood.berkeley.edu/article/item/four_ways_sadness_may_be_good_for_you.
8. Ad Vingerhoets. *Why Only Humans Cry: Unravelling the Mysteries of Tears* (Oxford University Press, 2013).
9. Rachel Garner, "Elephants Don't Have Tear Ducts—so Why Are They Always Crying?," Why Animals Do the Thing, June 9, 2018, https://www.whyanimalsdothething.com/elephants-dont-cry.
10. William H. Frey, Denise DeSota-Johnson, Carrie Hoffman, and John T. McCall, "Effect of Stimulus on the Chemical Composition of Human Tears," *American Journal of Ophthalmology* 92 (1981): 559–67.
11. Lauren M. Bylsma, Ad J. J. M. Vingerhoets, and Jonathan Rottenberg, "When Is Crying Cathartic? An International Study," *Journal of Social and Clinical Psychology* 27, no. 10, (2008): 1165–87.
12. Dalbir Bindra, "Weeping: A Problem of Many Facets," *Bulletin of the British Psychological Society* 25 (1972): 281–84.
13. James J. Gross, Barbara L. Frederickson, and Robert W. Levenson, "The Psychophysiology of Crying," *Psychophysiology* 31, no. 5 (1994): 460–68.
14. See Jonathan Rottenberg, Lauren M. Bylsma, and Ad J. J. M. Vingerhoets, "Is Crying Beneficial?," *Current Direction in Psychological Science* 17, no. 6 (2008): 400–404. This study is a review of the role of context (lab versus the field) in crying and proposes how to study crying.
15. Rottenberg et al., "Is Crying Beneficial?"
16. Michelle C. P. Hendriks, Marcel A. Croon, and Ad J. J. M. Vingerhoets, "Social Reactions to Adult Crying: The Help-Soliciting Function of Tears," *Journal of Social Psychology* 148, no. 1 (2008): 22–42.
17. George Bonanno, *The Other Side of Sadness: What the New Science of Bereavement Tells Us About Life After Loss* (Basic, 2019), 42.

18. Christine D. Wilson-Mendenhall, Lisa Feldman Barrett, and Lawrence W. Barsalou, "Neural Evidence That Human Emotions Share Core Affective Properties," *Psychological Science* 24, no. 6 (2013): 947–56.
19. Jie Li, Magaret Stroebe, Cecilia L. W. Chan, and Amy Y. M. Chow, "The Bereavement Guilt Scale: Development and Primary Validation," *Omega: Journal of Death and Dying* 75, no. 2 (2017): 166–83. See also Rajendra A. Morey, Gregory McCarthy, Elizabeth S. Selgrade, Sristi Seth, Jessica D. Nasser, and Kevin LaBar, "Neural Systems for Guilt from Actions Affecting Self Versus Others," *Neuroimage* 60, no. 1 (2012): 683–92.
20. Margaret S. Miles and Alice S. Demi, "Toward the Development of a Theory of Bereavement Guilt: Sources of Guilt in Bereaved Parents," *Omega: Journal of Death and Dying* 14, no. 4 (1984): 299–314.
21. Phillip R. Shaver and Caroline M. Tancredy, "Emotion, Attachment, and Bereavement: A Conceptual Commentary," in *The Handbook of Bereavement Research: Consequences, Coping, and Care*, ed. Margaret S. Stroebe, Robert O. Hansson, Wolfgang Stroebe, and Henk Schut (American Psychological Association, 2001), 63–88.
22. Bonanno, *The Other Side of Sadness*, 59.
23. George A. Bonanno, Laura Goorin, and Karin G. Coifman, "Sadness and Grief," in *The Handbook of Emotions*, 3rd ed., ed. Michael Lewis, Jeannette M. Haviland-Jones, and Lisa Feldman Barrett (Guilford Press, 2008), 797–810; George A. Bonanno, "Grief and Emotion: A Social-Functional Perspective," in *The Handbook of Bereavement Research*, ed. Stroebe et al., 493–515.

5. PAIN

1. Wong-Baker FACES Foundation, "Wong-Baker FACES History," https://wongbakerfaces.org/us/wong-baker-faces-history/, accessed February 7, 2025.
2. David Biro, "Is There Such a Thing as Psychological Pain? And Why It Matters," *Culture, Medicine, and Psychiatry* 34 (2010): 658–67.
3. Ronald Melzack and Patrick D. Wall, *The Challenge of Pain*, 2nd ed. (Penguin Press, 1988).
4. Miriam Stoeber, Damien Jullié, Braden T. Lobingier, Toon Laeremans, Jan Steyaert, Peter W. Schiller, et al., "A Genetically Encoded

Biosensor Reveals Location Bias of Opioid Drug Action," *Neuron* 98 (2018): 963–76.

5. Gun-Yeon Na, "Evaluation of the Life Span in Leprosy Patients," *Korean Journal of Dermatology* 32, no. 1 (1994): 8–12.
6. Samantha K. Brooks, Rebecca K. Webster, Louise E. Smith, Lisa Woodland, Simon Wessely, Neil Greenberg, et al., "The Psychological Impact of Quarantine and How to Reduce It: Rapid Review of the Evidence," *Lancet* 395 (2020): 912–20.
7. Saachi Arora and Sangeeta Bhatia, "Addressing Grief and Bereavement in Covid-19 Pandemic," *Illness, Crisis, and Loss* 31, no. 1 (2023): 1–12.
8. Debra Jackson and Kim Usher, "Understanding Expressions of Public Grief: 'Mourning Sickness,' 'Grief-lite,' or Something More?," *International Journal of Mental Health Nursing* 24 (2015): 93–94.
9. William James, *The Principles of Psychology*, vol. 1 (1890) (Dover Press, 1950).
10. Zhansheng Chen, Kipling D. Williams, Julie Fitness, and Nicola C. Newton, "When Hurt Will Not Heal: Exploring the Capacity to Relive Social and Physical Pain," *Psychological Science* 19, no. 8 (2008): 789–95.
11. Donald D. Price, "The Psychological and Neural Mechanisms of the Affective Dimension of Pain," *Science* 288, no. 5472 (2000): 1969–72.
12. Phillip Gerrans, "Pain Asymbolia as Depersonalization for Pain Experience: An Interoceptive Active Inference Account," *Frontiers in Psychology* 11 (October 2020): art. 523710, https://doi.org/10.3389 fpsyg.2020.523710.
13. Naomi I. Eisenberger, Tristen K. Inagaki, Nehjla M. Mashal, and Michael R. Irwin, "Inflammation and Social Experience: An Inflammatory Challenge Induces Feelings of Social Disconnection in Addition to Depressed Mood," *Brain, Behavior, and Immunity* 24, no. 4 (2010): 558–63.
14. See, for example, Naomi I. Eisenberger, Johanna M. Jarcho, Matthew D. Lieberman, and Bruce D. Naliboff, "An Experimental Study of Shared Sensitivity to Physical Pain and Social Rejection," *Pain* 126, nos. 1–3 (2006): 132–38.
15. C. Nathan DeWall, Geoff Macdonald, Gregory D. Webster, Carrie L. Masten, Roy F. Baumeister, Caitlin Powell, et al., "Acetaminophen Reduces Social Pain: Behavioral and Neural Evidence," *Psychological Science* 21 (2010): 931–37.

16. Mani Pavuluri and Amber May, "I Feel, Therefore, I Am: The Insula and Its Role in Human Emotion, Cognition, and the Sensory-Motor System," *AIMS Neuroscience* 2, no. 1 (2015): 18–27.
17. Naomi I. Eisenberger, "Broken Hearts and Broken Bones: A Neural Perspective on the Similarities Between Social and Physical Pain," *Current Directions in Psychological Science* 21, no. 1 (2012): 42–47.
18. Barna Konkoly Thege, Janos Pilling, Zoltan Cserhati, and Maria S. Kopp, "Mediators Between Bereavement and Somatic Symptoms," *British Medical Journal Family Practice* 13 (2012): art. 59, http://www.biomedcentral.com/1471-2296/13/59.

6. GRIT, RESILIENCE, AND GRACE

1. Kai Epstude and Neal J. Roese, "The Functional Theory of Counterfactual Thinking," *Personality and Social Psychology Review* 12, no. 2 (2008): 168–92.
2. Maarten C. Eisma and Margaret S. Stroebe, "Rumination Following Bereavement: An Overview," *Bereavement Care* 36, no. 2 (2017): 58–64, https://doi.org/10.1080/02682621.2017.1349291.
3. Jeannett M. Smith and Lauren B. Alloy, "A Roadmap to Rumination: A Review of the Definition, Assessment, and Conceptualization of This Multifaceted Construct," *Clinical Psychology Review* 29, no. 2 (2009): 116–28, https://doi.org/10.1016/j.cpr.2008.10.003.
4. Maarten C. Eisma, Paul A. Boelen, Henk A. W. Schut, and Margaret S. Stroebe, "Does Worry Affect Adjustment to Bereavement? A Longitudinal Investigation," *Anxiety, Stress, and Coping* 30, no. 3 (2017): 243–52, https://doi.org/10.1080/10615806.2016.1229464.
5. Margaret S. Stroebe and Henk Schut, "The Dual Process Model of Coping with Bereavement: Rationale and Description," *Death Studies* 23, no. 3 (1999): 197–224.
6. Eisma et al., "Does Worry Affect Adjustment to Bereavement?," 244.
7. Kimberly A. Calderwood, "Adapting the Transtheoretical Model of Change to the Bereavement Process," *Social Work* 56, no. 2 (2011): 107–18.
8. Calderwood, "Adapting the Transtheoretical Model of Change to the Bereavement Process," 110.

9. Phillipa Lally, Cornelia H. M. Van Jaarsveld, Henry W. W. Potts, and Jane Wardle, "How Habits Are Formed: Modelling Habit Formation in the Real World," *European Journal of Social Psychology* 40 (2010): 998–1009.
10. R. Nicholas Carleton, "Fear of the Unknown: One Fear to Rule Them All?," *Journal of Anxiety Disorders* 41 (2016): 5–21.
11. C. S. Lewis, *A Grief Observed* (1961; reprint, Harper Collins, 1996), 3.
12. Samantha Stein, "Grief and Fear," *Psychology Today*, September 26, 2015, https://www.Psychologytoday.com/us/blog/what-the-wild-things-are/201509/Grief-and-fear.
13. Peter J. Freed, Ted K. Yanagihara, Joy Hirsch, and J. John Mann, "Neural Mechanisms of Grief Regulation," *Biological Psychiatry* 66, no. 1 (2009): 33–40, https://doi.org/10.1016/j.biopsych.2009.01.019.
14. Mary-Francis O'Connor, David K. Wellisch, Annette L. Stanton, Naomi I. Eisenberger, Michael R. Irwin, and Matthew D. Liberman, "Craving Love? Enduring Grief Activates Brain's Reward Center," *Neuroimage* 42 (2008): 969–72.
15. Adriana Feder, Sharely Fred-Torres, Steven M. Southwick, and Dennis S. Charney, "The Biology of Human Resilience: Opportunities for Enhancing Resilience Across the Life Span," *Biological Psychiatry* 86 (2019): 443–53.
16. Steven M. Southwick, George A. Bonanno, Ann S. Masten, Catherine Panter-Brick, and Rachel Yehuda, "Resilience Definitions, Theory, and Challenges: Interdisciplinary Perspectives," *European Journal of Psychotraumatology* 5 (2014): art. 25338, https://doi.org/10.3402/ejpt.v5.25338.
17. "Executive Function," Weil Institute for Neurosciences, University of California at San Francisco, https://memory.ucsf.edu/symptoms/executive-functions#:~:text=The%20executive%20system%20involves%20the,our%20closest%20nonhuman%20primate%20relatives, accessed November 2, 2022.
18. Steven F. Maier and Linda R. Watkins, "Role of the Medial Prefrontal Cortex in Coping and Resilience," *Brain Research* 1355 (2010): 52–60, https://doi.org/10.1016/j.brainres.2010.08.039.
19. Stephan Maul, Ina Giegling, Chiara Fabbri, Filippo Corponi, Alessandro Serretti, and Dan Rujescu, "Genetics of Resilience: Implications for

Genome-Wide Association Studies and Candidate Genes of the Stress Response System in Posttraumatic Stress Disorder and Depression," *American Journal of Medical Genetics* 183B (2020): 77–94.

20. *APA Dictionary of Psychology*, s.v. "grit, *n.*," updated April 19, 2018, https://dictionary.apa.org/grit.
21. Kristin Meekhof, "Grit, Grief, and Grace: How One Woman Managed Loss," *Psychology Today*, July 12, 2018, https://www.psychologytoday.com/us/blog/widows-guide-healing/201807/grief-grit-and-grace.
22. Anne Lamott, quoted in Tyler Kleeberger, "Why We Resist Change," Medium, February 25, 2021, https://medium.com/becominghuman-tylerkleeberger/why-we-resist-change-3c5ba7d8d698.
23. J. W. Pennebaker and S. K. Beall, "Confronting a Traumatic Event: Toward an Understanding of Inhibition and Disease," *Journal of Abnormal Psychology* 95, no. 3 (1986): 274–81.
24. J. M. Smyth, "Written Emotional Expression: Effect Sizes, Outcome Types, and Moderating Variables," *Journal of Consulting and Clinical Psychology* 66 (1998): 174–84.
25. D. M. Sloane and B. P. Marx, "Taking Pen to Hand: Evaluating Theories Underlying the Written Disclosure Paradigm," *Clinical Psychology: Science and Practice* 11, no. 2 (2004): 121–37.

7. FINDING MEANING

1. David E. Balk, "A Modest Proposal About Bereavement and Recovery" (literature review), *Death Studies* 32 (2008): 84–93; Kimberly A. Calderwood, "Adapting the Transtheoretical Model of Change to the Bereavement Process," *Social Work* 56, no. 2 (2011): 107–18.
2. George Hagman, "Beyond Decathexis: Toward a New Psychoanalytic Understanding and Treatment of Mourning," in *Meaning Reconstruction and the Experience of Loss*, ed. Robert A. Neimeyer, Kindle ed. (American Psychological Association, 2001), chap. 1.
3. Thomas Attig, "Relearning the World: Making and Finding Meaning," in *Meaning Reconstruction and the Experience of Loss*, ed. Neimeyer, loc. 802, emphasis in original.
4. *APA Dictionary of Psychology*, s.v. "narrative psychology, *n.*," updated April 19, 2018, https://dictionary.apa.org/narrative-psychology.

5. Robert A. Neimeyer, "Introduction: Meaning Reconstruction and Loss," in *Meaning Reconstruction and the Experience of Loss*, ed. Neimeyer, loc. 185.
6. Robert A. Neimeyer, Dennis Klass, and Michael Robert Dennis, "A Social Constructionist Account of Grief: Loss and the Narration of Meaning," *Death Studies* 38, nos. 6–10 (2014): 486, https://doi.org/10.1080/07481187.2014.913454.
7. Rachel A. Coleman and Robert A. Neimeyer, "Measuring Meaning: Searching for and Making Sense of Spousal Loss in Late-Life," *Death Studies* 34, no. 9 (2010): 805, https://doi.org/10.1080/07481181003761625.
8. James Gillies and Robert A. Neimeyer, "Loss, Grief, and the Search for Significance: Toward a Model of Meaning Reconstruction in Bereavement," *Journal of Constructivist Psychology* 19 (2006): 31–65, https://doi.org/10.1080/10720530500311182.
9. Neimeyer et al., "A Social Constructivist Account of Grief," 490.
10. Thomas T. Frantz, Barbara C. Trolley, and Megan M. Farrell, "Positive Aspects of Grief," *Pastoral Psychology* 47, no. 1 (1998): 3–17.
11. Judith Mangelsdorf, Michael Eid, and Maike Luhmann, "Does Growth Require Suffering? A Systematic Review and Meta-Analysis on Genuine Posttraumatic and Postecstatic Growth," *Psychological Bulletin* 145, no. 3 (2019): 302–38.
12. Martin E. P. Seligman and Mihaly Csikszentmihalyi, "Positive Psychology: An Introduction," *American Psychologist* 55, no. 1 (2000): 5–14; James O. Pawelski, "William James, Positive Psychology, and Healthy-Mindedness," *Journal of Speculative Philosophy* 17, no. 1 (2003): 53–67; Jeffrey J. Froh, "The History of Positive Psychology: Truth Be Told," *NYS Psychologist* 16, no. 3 (2004): 18–20.
13. Carlos Laranjeira and Ana Querido, "Hope and Optimism as an Opportunity to Improve 'Positive Mental Health' Demand," *Frontiers in Psychology* 13 (2022): art. 827320, https://doi.org/10.3389/fpsyg.2022.827320.
14. Rama K. Rajandram, Samuel M. Y. Ho, Nabil Samman, Natalie Chan, Colman McGrath, and Roger A. Zwahlen, "Interaction of Hope and Optimism with Anxiety and Depression in a Specific Group of Cancer Survivors: A Preliminary Study," *BMC Research Notes* 4 (2011): art. 519, https://doi.org/10.1186/1756-0500-4-519.

15. Matthew W. Gallagher and Shane J. Lopez, "Positive Expectancies and Mental Health: Identifying the Unique Contributions of Hope and Optimism," *Journal of Positive Psychology* 4, no. 6 (2009): 548–56, https://doi.org/10.1080/1743976090; Liz Day, Katie Hanson, John Maltby, Carmel Proctor, and Alex Wood, "Hope Uniquely Predicts Objective Academic Achievement Above Intelligence, Personality, and Previous Academic Achievement," *Journal of Research in Personality* 44 (2010): 550–53.
16. Laranjeira and Querido, "Hope and Optimism as an Opportunity to Improve."
17. M. Katherine Shear, "Grief and Mourning Gone Awry: Pathway and Course of Complicated Grief," *Dialogues in Clinical Neuroscience* 14, no. 2 (2012): 119–28.
18. Song Wang, Xin Xu, Ming Zhou, Taolin Chen, Xun Yang, Guangiang Chen, et al., "Hope and the Brain: Trait Hope Mediates the Protective Role of Medial Orbitofrontal Cortex Spontaneous Activity Against Anxiety," *NeuroImage* 157 (2017): 439–47, https://doi.org/10.1016/j.neuroimage.2017.05.056.
19. See, for example, Mohammed R. Milad and Scott. L. Rauch, "The Role of the Orbitofrontal Cortex in Anxiety Disorders," *Annals of the New York Academy of Science* 1121 (2007): 546–61.
20. Wang et al., "Hope and the Brain."

8. STORIES OF GRIEF AND GRIEVING

1. U.S. Institute of Medicine, Committee for the Study of Health Consequences of the Stress of Bereavement, *Bereavement: Reactions, Consequences, and Care*, ed. Marian Osterweis, Fredric Solomon, and Morris Green (National Academy Press, 1984), chap. 4: "Reactions to Particular Types of Bereavement," full text available online at https://pubmed.ncbi.nlm.nih.gov/25032459//pdf/Bookshelf_NBK217842.pdf.
2. Colin Murray Parkes and Holly G. Prigerson, "Determinants of Grief—I. Kinship, Gender, and Age," chap. 9 in Parkes and Prigerson, *Bereavement: Studies of Grief in Adult Life*, 4th ed. (Routledge Press, 2010), 137–51.

3. Nancy E. Thacker, "Musical Mourning in Manchester: Cultural Norms, Expectation, and Meaning in Grief," *Death Studies* 46, no. 7 (2022): 1689–96.
4. Hospice Foundation, "Common Questions," https://hospicefoundation.org/common-questions/, accessed February 7, 2025.
5. A former student, Leah Owenby, and her mom, E. T. Mitchell, found that the quote comes from Anne Lamott, *Plan B: Further Thoughts on Faith* (Riverhead Books, 2005), 174.

INDEX